THE QUANTUM ADVENTURE

Does God Play Dice?

Alex Montwill • Ann Breslin

University College Dublin, Ireland

Imperial College Press

Imperial College Press
57 Shelton Street
Covent Garden
London WC2H 9HE

Distributed by

World Scientific Publishing Co. Pte. Ltd.

5 Toh Tuck Link, Singapore 596224

USA office: 27 Warren Street, Suite 401-402, Hackensack, NJ 07601

UK office: 57 Shelton Street, Covent Garden, London WC2H 9HE

British Library Cataloguing-in-Publication Data
A catalogue record for this book is available from the British Library.

THE QUANTUM ADVENTURE
Does God Play Dice?

Copyright © 2012 by Imperial College Press

ISBN-13 978-1-84816-647-9
ISBN-10 1-84816-647-8
ISBN-13 978-1-84816-648-6 (pbk)
ISBN-10 1-84816-648-6 (pbk)

Typeset by Stallion Press
Email: enquiries@stallionpress.com

Printed by FuIsland Offset Printing (S) Pte Ltd Singapore

This book is dedicated to Ann and Liam,
our life partners

Contents

Acknowledgements

We wish to thank Lorraine Hanlon, Head of the UCD School of Physics, for her support and for providing us with the facilities of the School. Our colleagues were always ready to help; in particular, Emma Sokell, who assisted with setting up the imaging of demonstration experiments and John Quinn, who was always there to rescue us from the vagaries of computers. John White made numerical calculations of graphs and interference patterns.

Special thanks are due to Bairbre Fox, and Marian Hanson who, at all times, were ready to be of assistance.

It was a pleasure to work with Tasha D'Cruz of Imperial College Press who guided us through the task with skill and patience.

We are indebted to all who gave us high resolution images of their experiments and copyright permissions. They are acknowledged individually in the text.

Introduction

Science in the Comfort Zone

At the birth of the 20th century, science was generally accepted to be in a satisfactory state of health. Generations of *natural philosophers* had built a coherent structure which stood on foundations laid 400 years earlier by Galileo Galilei and Isaac Newton.

One of the basic tenets of the scientific method was that the development of a physical system is determined completely by its previous state. There is a well-defined relationship between cause and effect and the physical laws give no 'freedom of choice'; identical experiments give identical results. This principle was taken as an axiom, self-evident and necessary for the very existence of science.

Within this paradigm Newton's laws explained the motion of the planets and even predicted the existence of an undiscovered planet. Thermodynamics described processes involving heat energy with great accuracy. Maxwell's equations led to a complete understanding of how electric and magnetic fields behaved; his ultimate triumph was to show how electromagnetic waves, such as light and radio waves, were produced. Such theories led to many practical applications such as steam engines, electrical power, wireless communication and even flying machines.

Planck's Adventure

The train of events described in this book was triggered by a startling discovery made by Max Planck; it led to an adventure of the mind, unprecedented in the history of physics. The participants were faced with an unexpected new world in which the 'normal' rules of logic and 'common sense' did not apply.

Planck, a Professor of Theoretical Physics at the University of Berlin, did not intentionally set out on an adventure. He was analysing data, taken at the Physikalisch-Technische Reichsanstalt during experiments designed to optimize the efficiency of lighting devices. In trying to understand the basic process of light emission from glowing hot metals, Planck was confronted with the perplexing situation that, according to the 'well-established' theories, the amount of energy emitted from a hot metallic surface should be infinite.

Something was very wrong. '*A solution had to be found at any price*'. In '*an act of desperation*', Planck proposed that energy is emitted, not smoothly, but in indivisible 'bundles' or *quanta*. This enabled him to construct a formula which gave excellent agreement with the experiment, but it replaced one conundrum with another. What kind of a law would limit physical processes in this fashion? Planck could give no reason why nature should impose such a constraint, other than that it enabled him to construct a formula which agreed with experimental results.

Einstein, Bohr and Born Come on Board

At first, most of Planck's peers did not take him seriously but there were exceptions, notably Albert Einstein, Niels Bohr and Max Born, who embraced the new idea with enthusiasm.

Einstein quickly remodelled Planck's quantum of energy into a particle-like bundle which hurtles through space like an atomic billiard ball, capable of knocking electrons right out of metals.

Bohr introduced the quantum into the planetary model of the atom, providing an explanation for the characteristic appearance of atomic spectra. He went on to create an institute in Copenhagen which was devoted solely to the study of quantum physics. Bohr's mission was to bring together the brightest young scientists from all over the world. It was an unqualified success; Bohr and his young colleagues developed what became known as the 'Copenhagen Interpretation' of quantum mechanics.

Max Born established a second centre for quantum studies at the Institute of Theoretical Physics in Göttingen, which provided additional mathematical back-up to the more philosophical deliberations in Copenhagen.

The Future is a Lottery?

It soon became clear that Planck's 'act of desperation' did much more than introduce a radical new concept regarding the emission of energy. It had knock-on effects which led to far-reaching departures from well-established scientific principles. One of these effects was to introduce spontaneous 'quantum jumps' between 'allowed' energies in the atom. These were assumed to be unpredictable, which meant that atomic phenomena do not follow a predetermined path but are governed by probability, a symbolic 'throw of the dice'.

Albert Einstein could not come to terms with this development. In his view, the scientific method demands that the same initial conditions must always lead to the same sequence of results. One of the rock-solid foundation stones of science was under threat; if that was removed the whole structure would collapse. He could not accept that the future of the universe is determined by successive arbitrary events at the atomic level: '*Der Alte würfelt nicht: God (the Old One) does not throw dice*'.

Erwin Schrödinger, another central figure in the development of the theory, joined Einstein in the camp of the objectors. To demonstrate the absurdity of what he had helped to create, he expressed his doubts in the paradox of *Schrödinger's cat*, which is mysteriously both alive and dead. That same year, Einstein, together with Boris Podolsky and Nathan Rosen came up with an even more subtle paradox, involving '*spooky actions at a distance*'.

The Adventure Continues

Over a century after the quantum adventure began, experiments are verifying quantum phenomena which defy understanding. There may not be cats which are simultaneously both alive and dead, but in the world of the atom such 'multiple existence' is commonplace. Experiments with electrons show 'spooky entanglement at a distance'; the electrons are intertwined regardless of how far apart they are. Light not only hurtles through space like a billiard ball but, at other times, behaves as if it is controlled by the throw of a dice. Apparently nature does not recognize 'perfectly reasonable' axioms.

We set out to tell the story of this adventure and the people who took part in it. It describes a journey into a world we cannot directly experience and which seems to be a world of fiction; '*Curiouser and curiouser!*', as Alice says in Lewis Carroll's book, *Alice in Wonderland*.

We will try to persuade the reader that the quantum mechanical world is not some sort of fictional wonderland but real; incredible only because it is different from what we think the world should be.

Richard Feynman expressed the dilemma in these words:

> *We always have had a great deal of difficulty in understanding the world view that Quantum Mechanics represents. At least I do, because I'm an old enough man that I haven't got to the point that this stuff is obvious to me ... It has not yet become obvious to me that there is no real problem. I cannot define the real problem, therefore I suspect that there is no real problem, but I'm not sure that there's no real problem.*[1]

---

[1] A.J.G. Hey (ed.). *Feynman and Computation: Exploring the Limits of Computers.* Perseus Books, Reading, Mass. 1998.

Chapter 1

Prehistory — Isaac Newton

Before the advent of the quantum adventure, physical phenomena were described and understood, or thought to be understood, by the laws of *classical mechanics*.

Nicolaus Copernicus (1473–1543)

The first step towards a theory of classical mechanics came in 1543, when a book entitled *De Revolutionibus Orbium Coelestium* was published. It created a great deal of controversy but its author Nicolaus Copernicus[1] barely lived to see the final printed version, and was spared the hostile reaction to his revolutionary ideas. What created the stir was his assertion that the Earth and the other planets are orbiting the sun. It was a simple and elegant description of celestial motion, and was backed by observation and detailed mathematical calculation. What upset his critics was the notion that the

Nicolaus Copernicus. Courtesy of Poczta Polska. © alexsol

Earth was not the centre of the universe. Only the moon revolves around us, not the sun and the stars.

---

[1] This is the Latin version of his Polish name 'Mikołaj Kopernik'.

Galileo Galilei (1564–1642)

Galileo Galilei. Courtesy of An Post, Irish Post Office.

Galileo pioneered the application of mathematical argument to physical theory. He was a strong supporter of the Copernican model, which was still strongly opposed by followers of the *Ptolemaic* or 'fixed Earth' system. How could the Earth be moving when we do not experience any effect of motion? The 'Ptolomaics' even came up with a physical argument that a stone dropped from a great height such as the Tower of Pisa is seen to fall vertically in a straight line, not at an angle due to the motion of the Earth. Galileo answered with an argument which was well ahead of its time. In his own words:

> *Shut yourself up with some friend in the largest room below decks of some large ship, and there procure gnats, flies, and such other small winged creatures. Also a great tub full of water and within it put certain fishes; let also a certain bottle be hung up, which drop by drop lets forth its water into another narrow-necked bottle placed directly underneath. Having observed how the small winged animals fly with like velocities towards all parts of the room, how the fishes swim indifferently towards all sides, and how the distilling drops all fall into the bottle placed underneath ... you will not be able to discern the least alteration in all the above-mentioned effects, or gather by any of them whether the ship moves or stands still.*

Nowadays Galileo's argument can be made even more impressive by referring to what happens in the cabin of a jet plane. The passengers carry on normal activities as long as the jet is flying smoothly at constant speed. Only while the plane accelerates at take off or slows down on landing, or passes through turbulence, do they feel any effects of motion and need to fasten seat belts. Galileo fell short of stating explicitly that an object in uniform motion, protected from any outside influences, will continue that motion forever. He may very well have understood that to be the case.

Isaac Newton (1642–1727)

Isaac Newton, born in the year that Galileo died, was the first to formulate Galileo's ideas in a quantitative mathematical way. His famous work *Philosophiae Naturalis Principia Mathematica*, published in 1686, marked the birth of classical mechanics and is arguably the most important book ever published in the history of science. Newton's basic principle was that things that are moving tend to keep moving and things that are stationary tend to remain stationary. No object has the ability either to move itself or stop itself.

Newton combined an amazing vision with an extraordinary talent for mathematics. He went about his work in a logical way, carefully selecting the relevant quantities relating to motion and developing deceptively simple mathematical relationships between them.

New concepts — momentum and force

Newton's first step was to introduce a new quantity called *momentum*, which was defined as *mass* multiplied by *velocity*:

$$\text{Momentum, } p = mv.$$

He then defined *force* as that which changes momentum and made what was to prove his most powerful quantitative statement that '*the rate at which momentum changes is a measure of that force*'.

Basically, Newton was saying that we need forces to set objects in motion and forces to slow them down, and that heavier objects need larger forces. These facts were well known and had been employed in a practical sense since time immemorial. What Newton did was to establish the precise nature of the relationship between force and movement.

A new idea — action and reaction

When a battering ram is smashed into a door, it is obvious that a force must be applied to stop it. Newton realized that at every impact, not one but *two* forces act. They are equal in size and opposite in direction (*action* and *reaction*) — the door stops the battering ram while the battering ram breaks down the door!

Battering ram

Newton's laws

Newton summarized his theory of motion in three laws:

AXIOMATA SIVE LEGES MOTUS

Lex I

Corpus omne perseverare in statu suo quiescendi vel movendi uniformiter in directum, nisi quatenus illud a viribus impressis cogitur statum suum mutare.

Lex II

Mutationem motus proportionalem esse vi motrici impressæ, & fieri secundum lineam rectam qua vis illa imprimitur.

Lex III

Actioni contrariam semper & æqualem esse reactionem: sive corporum duorum actiones in se mutuo semper esse æquales & in partes contrarias dirigi.

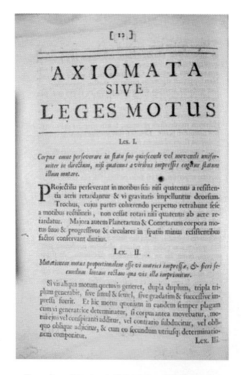

Page from Newton's Principia, 1687 edition

Law 1. *A body remains at rest or uniform motion in a straight line, unless it is acted on by an external force.*

Law 2. *Force equals rate of change of momentum.*

Law 3. *To every action there is an equal and opposite reaction.*

Work and energy

The concepts of *work* and *energy* are based on Newton's second law. If we want to increase the velocity of a particle, or push it up a hill, we have to apply a force, and that means we have to do work. For instance, to move a particle horizontally through a distance d, by applying a constant force F in the same direction, we need to do an amount of work, $W = Fd$.

This work is 'invested' in the particle and its energy increases by exactly that amount. This energy can be energy of motion (*kinetic* energy) or energy of configuration (*potential* energy) or a bit of both. In the case of a particle being pushed up a hill, its potential energy increases as it gains height, and if it moves faster as it climbs, its kinetic energy also increases.

In any event:

Total work done by the force
  = increase in potential energy + increase in kinetic energy.[2]

Changing the direction of motion

Motorbike taking a corner at speed. Courtesy of Robert Michael/AFP/Getty Images

Velocity and therefore momentum have both size and direction (they are *vectors*) and force must be applied to change the *direction* of the momentum of an object, even if it travels at constant speed.

Take the example of a motorbike racing around a track. There is no net force when the bike travels at constant velocity (in a straight line at constant speed). The force propelling the bike is matched by an equal and opposite force due to friction. However, we must supply additional force to bring it around a bend, even at constant speed. In fact, a number of forces combine to curb the tendency to continue in a straight line. These, in the main, are supplied by friction between the road surface and the tyres. The image above illustrates how the

---

[2] Energy can take other forms which do not concern us here.

rider is able to change the configuration of both himself and the bike, to generate the forces needed to change his direction of motion.

Newton applies his laws of motion to celestial objects

The moon orbits the Earth, held in place by the force of gravity.[3] Newton put forward the hypothesis, which at the time was new and original, that there is no difference between 'celestial' phenomena and phenomena on Earth. He assumed that the force which keeps the moon in orbit is exactly the same as the force which causes an apple to fall from a tree, the force of gravity.

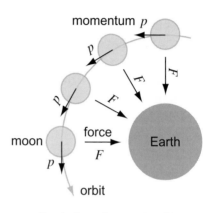

Gravity keeps the moon in orbit

Both the apple and the moon are falling towards the Earth. The difference is that the moon is in orbit and, while it falls, it does not get nearer!

This is another example of force producing a change in the direction of the momentum. The moon is like a celestial motorbike, being forced to travel in a continuous bend by gravity. The force between the Earth

and the moon acts along the line joining them and is always perpendicular to the direction the moon is moving; it can never change the *size* of the momentum (so the speed remains the same).

---

[3] More accurately stated, the moon and Earth orbit their common centre of mass.

Newton's law of gravitation

It is a tribute to Newton's genius that, quite independently from his laws of motion, he also formulated a *Universal Law of Gravitational Attraction*, which states that *any* two objects, *anywhere* in the universe attract one another. The force between them is proportional to their masses and inversely proportional to their separation. This force between masses is very small and normally hardly noticeable; it only plays a significant role if at least one object is massive (like the Earth).

Newton realized that he could use his equations of motion to calculate the time for the moon to complete one orbit of the Earth if he knew the size of the Earth's gravitational force at the location of the moon.

All the information (the distance from the Earth to the moon, the radius of the Earth and the acceleration of objects in free fall) was to hand. Newton calculated the period of the lunar orbit to be just less than four weeks, in good agreement with the observed value of 28.3 days.

Remarkably, the earliest estimates of the radius of the Earth and the distance from the Earth to the moon date back to in the 3rd century BC. We can only wonder at the skill and ingenuity of Eratosthenes (275–192 BC) and Aristarchus (310–230 BC), the Greek mathematicians and astronomers, who made these measurements.

Galileo demonstrated that all objects, whether heavy or light, fall at the same rate. (The story goes that he dropped simultaneously light and heavy balls from the top of the Leaning Tower of Pisa and found that they hit the ground at the same time.) He figured out the relationship between distance and time in free fall and measured the acceleration of free-falling objects to be 9.8 m/s^2.

A new era

Newton's laws of motion marked the beginning of a new era. They could be applied to problems involving forces and accelerations in all kinds of motion, in fact to every conceivable dynamical problem. Concepts such as energy and momentum were defined and quantified. Newton had shown that even the motion of celestial objects is governed by his laws, thus making the motions less mysterious.

Newton acknowledged the work of those who preceded him, probably Galileo in particular, with these words written in a letter to Robert Hooke, dated 5 February 1676:

> *If I have seen further, it is by standing on the shoulders of giants.*

Determinism

Newtonian mechanics is *deterministic,* which means that, in principle, the behaviour of any mechanical system can be predicted from the initial conditions. Newton's laws are definite and precise, and assert that if we know the position and momentum of every particle in the universe, we can, in principle, predict the future and reconstruct the past. In practice, this can be done only for simple systems, and then only approximately, because it is not possible to have complete knowledge of initial conditions.

A difficult shot?

A sequence of events during a game of snooker makes a good illustration. A player strikes the white cue ball which then collides with the red ball which he wishes to pot into one of the corner pockets. After the collision the cue ball rolls around the table, bouncing off the side cushions and off other balls. In principle, the final rest positions of the cue ball (and of the other balls) can be calculated from the initial conditions using Newton's laws. In practice one can make quite a good estimate of the result — expert snooker players make these calculations 'in their heads' — and try to leave the cue ball in a good position for the next shot. It is only an estimate, because frictional forces due to millions of random microscopic interactions generate cumulative effects as the number of collisions and reflections increases; it is not due to any uncertainty in Newton's laws.

Who invented calculus?

Newton needed mathematical methods to calculate velocities, accelerations, and generally rates of change in space and time. For this purpose he developed his method of *fluxions* and *fluents,* a technique now known as differential and integral calculus. The technique could also be applied to calculate maxima and minima, the tangent at any point on a curve and the area under a curve. He described his methods in unpublished manuscripts written between 1666 and 1671 which first appeared in print in 1704 as an appendix to his *Opticks.*

Gottfried Leibniz (1648–1716) developed similar methods which he published in 1684 under the title *Nova Methodus pro Maximis et Minimis Itemque Tangentibus.* This started a controversy. Newton accused Leibniz of plagiarism; Leibniz was highly indignant and wrote to the secretary of the Royal Society (of which Newton was President), stating that he had never heard of fluxions, and demanded an apology. Far from apologizing, the Society replied in even more aggressive language. In 1712 the Royal Society appointed a commission to investigate the affair; its report, not surprisingly, vindicated Newton. Today Newton and Leibniz are generally regarded as independent inventors of calculus.

Chapter 2

Preparing for Quantum Mechanics

When Newton died in 1727, he left as his legacy three laws of motion which form the basis of *classical mechanics*. The laws were found to apply wherever they could be tested by observation or experiment, not only to objects on Earth but also to 'celestial matter' such as the orbiting moon. While no adjustment or improvements were required, there remained two main challenges. The first challenge was conceptual rather than practical, and was to show that Newton's laws can be derived from basic principles such as economy, symmetry and the simplicity of nature; those same principles which the ancient philosophers believed were an absolute criterion for all natural phenomena. The second was to find general and mathematically elegant methods of applying the laws. This was done very successfully over the following century and led to what is now known as *generalized classical mechanics*.

Leading into the Quantum Adventure

These refinements to Newton's theory were closely interlinked, and generalized mechanics developed from basic criteria of economy and perfection of the natural laws. Even more importantly, these criteria apply to the then still-unexplored world of atoms and the methods of generalized mechanics serve as a perfect introduction to quantum mechanics.

Fermat's theorems

To find a link to basic principles, we go back to about 100 years before Newton, to the work of Pierre de Fermat (1601–1665), who was a lawyer and French government official. That was his day job; his hobby was mathematics. He is probably best known for his *'Last Theorem'*, which states that it is impossible to find integer values of x, y and z which satisfy the 'simple looking' equation $x^n + y^n = z^n$ when n is greater than two.

Pierre de Fermat. Creation et gravure Andre Lavergne. Courtesy of La Poste

Fermat seems to have enjoyed frustrating other mathematicians by stating results and theorems without proof, and challenging them to find a solution. Fermat's last theorem was scribbled in the margin of his copy of the classic Greek text *Arithmetic*, together with the words: '*I have a truly marvellous demonstration of this proposition which this margin is too narrow to contain*'.[1]

No one else could solve the problem, and a prize of 100,000 marks was lodged at the Academy of Sciences in Göttingen, for anyone who could find a complete proof of Fermat's theorem for all values of n. It was not until 1997 that the English mathematician, Andrew Wiles, published a complete proof; it represented 11 years of work and involved 60 pages of mathematics. It seems that Fermat was correct but his 'proof' may have been wrong.

Refraction

Fermat was also interested in mathematical ways of finding *maxima* and *minima* and invented techniques similar to *calculus*, discovered

---

[1] Harold M. Edwards. *A Genetic Introduction to Algebraic Number Theory*. Springer, Berlin. 1996.

later by Newton and Leibniz. He applied these techniques to the phenomenon of *refraction*, the change of direction of light as it passes between different transparent materials. Refraction had been described almost 700 years earlier by Ibn Al-Haitham (965–1040), the 'father of modern optics' in his book, *Kitab Al-Manathr.*

In 1621, Willebrord Snell (1580–1626) made systematic measurements of this change of direction, and found that it depends on the angle at which light strikes the boundary between the materials. He showed that the ratio of the sine of that angle to the sine of the angle at which the light leaves, is constant. This constant is known as the *refractive index* for light passing between the two materials.

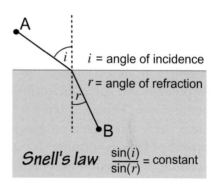

Snell's law

Snell gave no reason for his formula. He was not in a position to speculate why the ratio was constant or why it involved the sines and not for instance the cosines or tangents or the angles involved. He could not have known that it had any connection with the speed of light. At that time, even the question of whether light propagates at finite or infinite speed had not been answered.

The wave theory of light, developed about 200 years later, shows that Snell's law is a consequence of a change in the speed of propagation of the wave. The clue is there in the way ocean waves often change direction as they approach the shore and the water becomes more shallow. Waves travel more slowly in shallow than in deep water. The diagram shows what happens when a train of waves approaches a sandbank where the depth of water decreases

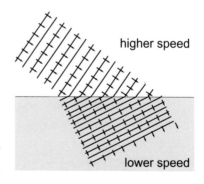

Waves change direction

abruptly. Each wavecrest swerves around because the part to the left slows down before the remainder. This change of direction progresses along the entire length of the wavefront, which then continues on in a new direction.

Fermat's principle of least time

Fermat took a global view. What has become known as *Fermat's principle of least time* states that when light travels from one point to another, it always chooses *the route which takes the shortest time*. His hypothesis was based on intuition and the belief in the perfection of natural laws. Fermat could not have known the speed of light; in his time there was a general acceptance of Aristotle's dogma that the speed of light is infinite.[2] It was a mathematical exercise using his technique of maxima and minima which, to his surprise and delight, gave results in agreement with Snell's law of refraction. Fermat deduced that light does take the quickest route and that what Snell called the refractive constant is the ratio of the speed of light in the first medium to the speed in the second medium.

What to us may be even more surprising is that Fermat's work was strongly criticized on 'philosophical' grounds. An example of such objections is found in a letter from Claude Clerselier (1614–1684), who makes a long and involved argument which includes the statement:

> *The principle that nature always acts by the shortest and simplest path, is only a moral principle, not a physical one — it is not and can not be the cause of any effect in nature...*

In a reply dated 21 May 1662 Fermat tries to dispel Clerselier's concerns:

> *I neither claim nor have ever claimed to be in the secret confidence of nature ... I have only offered her a small help on the subject of refraction,*

---

[2] The first scientific determination of the speed of light was made in 1676 by the Danish astronomer Olaus Römer (1644–1710).

if she ever to have need of it ... But since you assure me Sir, that she can manage her affairs without ... It is enough for me that you leave me in possession of my geometry problem, completely pure and in abstracto, by means of which one can find the path of a moving body which passes through two different media and which tries to complete its movement as soon as it can...

The shortest path is not always the quickest, as is illustrated in the photograph of light passing through a thick piece of glass. The light 'zig-zags' its way through the glass rather than going in a straight line, which would be the shortest route.

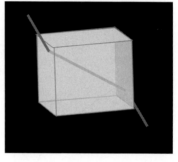

Refraction

Least time and lenses

The focussing of light by a lens is a simple, yet striking example of the principle of least time in action. We can trace the rays of light individually as they emerge from the source at A in all directions; those which enter the lens are refracted at the first surface as we can see in the diagram. They are

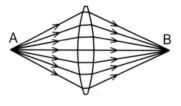

A choice of routes all equally quick

refracted again as they leave the lens at the second surface. If the lens is shaped correctly, these rays converge to a second point B. The lens thus fulfils its function of intercepting light rays diverging from A and bending them back by refraction to meet again at B.

Instead of meticulously following each individual ray, Fermat makes the grand statement that light follows the quickest route from A to B. In this case, all routes are equally quick. The straight line through the centre is the shortest but it also leads through the thickest part of the lens, where light travels more slowly than in air. The other routes are longer but the light passes through a smaller thickness of glass.

Least time and the lifeguard

The quickest path between two points is critical for a lifeguard who sees a swimmer in difficulty. His aim is to get to the swimmer in the shortest possible time. We can reasonably assume that the lifeguard can run faster than he can swim so he has to figure out how much time he should spend running, before entering the water.

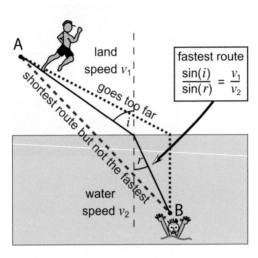

Lifeguard's quickest route

The diagram shows three possible routes that a lifeguard, starting from a point A, can take to get to a swimmer at point B.

The optimal route is a compromise between the straight line from A to B, where the segment at the lower speed is too long, and the more roundabout route where the lifeguard runs at the higher speed to the point nearest the swimmer before he enters the water. It turns out that the fastest route is the one which satisfies Snell's law.

A deeper insight into the laws of motion

Fermat's law applies specifically to the path taken by light, which raises the fascinating question: 'Can Newton's laws be derived from a similar universal principle which applies to all natural phenomena?' The first published statement[3] of such a principle is generally attributed to Pierre Louis Maupertuis (1698–1759), whose hypothesis that *'Nature is thrifty in all its actions'* was published in 1744. Not unlike Fermat, he based his assertion on an instinctive feeling that the perfection of nature demanded a certain economy,

---

[3] There is some (disputed) evidence that the concept was mentioned by Leibniz, in private correspondence with the mathematician Jakob Hermann (1678–1733).

as opposed to needless expenditure of energy, but he was unable to provide a quantitative argument to support it.

The principle of least action

Leonhard Euler (1707–1783), Joseph Louis Lagrange (1736–1813) and Jean Le Rond D'Alembert (1717–1783) all contributed to the development of a mathematical formalism to express Maupertius' notion of nature's thrift. They quantified 'economy' in terms of the difference between energy due to motion (kinetic energy) and energy due to position (potential energy). This somewhat mysterious function became known as the *Lagrangian*.

Lagrangian, $L = $ (kinetic energy – potential energy).

The Lagrangian function has a remarkable property: its average value for a physical system, as it evolves from one state to another, is smaller for the *actual path* taken by the system than for any other path. This is known as the *principle of least action*.

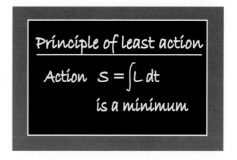

To calculate the average value of the Lagrangian as a system evolves, we measure the difference between the kinetic and potential energies of the system at regular time intervals and add the results. This sum is called the *action*. The formula for calculating the action appears on the blackboard.

Generalized mechanics

A new approach to mechanics grew out of the principle of least action. Euler in his book *Mechanica*, published in 1737, and D'Alembert in *Traité de Dynamique* (1743) independently developed a uniform analytic approach to mechanics. In Newton's method, every single stage of a process had to be treated individually. In the new approach, the goal was to deduce all physical phenomena by rigorous mathematical methods, based on simple principles, and to show that the laws of motion were logical necessities. Things could just not be any other way!

Lagrange summarized the 'new' mechanics in his book *Mécanique Analytique*, published in 1788; this became the authoritative work on classical mechanics. In the preface, Lagrange eloquently describes the purpose of the work:

> *To reduce the theory of mechanics and the art of solving associated problems, to general formulae, whose simple development provides all the equations necessary for the solution of each problem … To unite, and present from one point of view, the different principles which have so far been found to assist in the solution of problems in mechanics … No diagrams will be found in this work. The methods which I explain in it require neither constructions nor geometrical or mechanical arguments, but only the algebraic operations inherent to a regular and uniform process.*

The principle of least action at work

We can start with the simplest case where there are no external forces. Newton's first law states that a moving object will continue to move at constant velocity. Can we come to the same conclusion from the principle of least action?

If there are no forces, the potential energy is zero everywhere and the action depends only on the behaviour of the kinetic energy. When an object moves from one point to another in a fixed time, we can deduce that the action is minimal if its kinetic energy, and therefore its speed, is constant for the entire journey.

If the object goes faster than average for part of the journey and slower than average for the remainder, it could still manage to arrive at the same time as before, provided the faster bits make up exactly for the time lost during the slower portions. However, the kinetic energy depends on the *square* of the speed, so the object uses more energy when it goes faster than it saves by going more slowly i.e. the action is greater than if its speed were constant for the entire journey.

Least action on the track — Eddie and Fred

The principle of least action is familiar to runners. Let us try an example with numbers: Eddie and Fred set off together on a 6 km run which they intend to complete in one hour.

Clever Eddie, who knows about the principle of least action, jogs at a steady speed of 6 km/hr for the entire time, and arrives at the destination at exactly the appointed time.

Fred, on the other hand, walks at 4 km/hr for 3/4 hour and then realizes that he has another 3 km to complete. He then runs at 12 km/hr for the remaining 1/4 hour. He arrives at the finish at exactly the same time as Eddie so both of them averaged the same speed.

Which runner follows the course of least action? If we assume both of them have the same mass, we need compare only the numerical values of the average of the square of their speed:

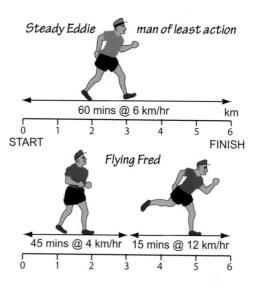

Eddie: average (speed)^2 = 6^2 = 36
Fred: average (speed)^2
        = 4^2 × (3/4) + 12^2 × (1/4)
        = 12 + 36 = 48

Fred's action is 33.3% greater than Eddie's. His catch-up run cost him as much as Eddie's action for the entire journey.

Distance runners are well aware of the advantages of maintaining a steady pace.

Emil Zátopek. *Courtesy of Popperfoto/Getty Images*

Emil Zátopek, one of the greatest athletes of his era, won three gold medals for Czechoslovakia at the 1952 Olympic Games in Helsinki. With his great power he used the tactic of changing pace at random intervals, making it all the more difficult for the other runners to keep up with him in a race.

Least action when there is a force at work

We can choose a more challenging example, such as the motion of a stone thrown into the air. Gravity causes the stone to lose speed[4] as it rises, momentarily come to rest, and then falls back towards the Earth.

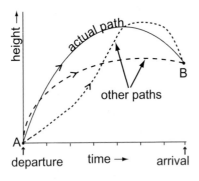

Possible paths of a stone thrown from A to B

As the stone gains height its potential energy increases, and at the same time, its kinetic energy decreases. After it has reached the highest point of its trajectory, the process reverses. It turns out that if you take the difference between the two kinds of energy at regular intervals and calculate the average value for the entire journey, this average will be smaller for the actual path than for any other path.

At the start, when the potential energy is zero, the stone travels fast to increase its potential energy as quickly as possible and reduce the Lagrangian, i.e. the *difference* between the potential and kinetic energy. However, it can't go *too* fast because it will then have too much kinetic energy and the difference will increase again. There is a balance between kinetic and potential energies throughout the journey. We can try any other path — higher, lower, even a straight line — but we will find the average is always greater.

In this example, it is much more difficult to find the actual path using least action than in the 'Eddie and Fred' example where the path was known. We would need to calculate the action for *all possible* paths

---

[4] More precisely, the vertical component of its velocity.

from A to B to find the one and only path which will take the stone from A to B in the specified time. A more sophisticated approach is needed.

The problem of finding the path *analytically*, on the basis of least action, was addressed by Lagrange and his peers. They translated the principle of least action into a set of equations which can be solved to reveal details of the motion of a particle or system of particles. Developed in the 1750s, they are known as the *Euler–Lagrange* equations and can be generalized to apply to any number of particles, with both external and internal forces acting on them, in fact to any system no matter how complex.

What is so different about least action?

The law of least action makes a statement about the whole evolution of a physical system, whereas Newton's laws tell us what happens at an individual point.

In the last example of the stone, Newton's laws apply continuously as the stone inches its way from point to point while the law of least action makes a grand statement about the entire process from beginning to end.

The really strange thing is that the stone seems to 'know' which is the path of least action. We might well ask: 'Does it explore all other possible paths and then choose the right one? Is the path determined with absolute certainty or could the stone choose another path  that is very close?' These questions may seem absurd and pointless for something like a stone, but what about subatomic particles such as electrons?

In Lagrange's time it was too early to address such questions, or even to speculate about what might be. The electron had not even been discovered, much less other subatomic particles! Future discoveries would show that questions which are absurd and effects which are negligible in the 'everyday world' become dominant in the world of atoms. In the quantum world, indeterminacy takes on a central role; certainty is replaced by probability. The Quantum Adventure had not yet begun.

Hamilton's method

In 1833 William Rowan Hamilton (1805–1865) developed the methods of Lagrange and formulated *Hamilton's equations*, which are easier to solve and have more physical meaning than Lagrange's equations. As a result, the principle of least action is often referred to as *Hamilton's principle*. However, Hamilton always paid tribute to Lagrange's work, describing the *Mécanique Analytique* as '*a scientific poem*'.

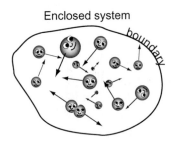

Enclosed system

Particles move within a boundary

The central feature of Hamilton's method is the *Hamiltonian function H* which expresses the total energy of a system in terms of the position, q and the momentum, p of every particle in it. These parameters completely describe the dynamics of the system.

The diagram shows a system of particles surrounded by a boundary. The particles are trapped inside, and nothing can cross the boundary in either direction — it is called an 'enclosed' system.

At any particular instant, we can describe the position of any particle in terms of its space coordinates (x, y, and z) and its motion in terms of the corresponding components of its momentum.[5] This gives us a total of six coordinates for each particle.[6]

Hamilton's equations of motion

The motion of each particle is described by six equations which describe how a change in any of the coordinates affects the

---

[5] In mathematical terms, momentum is treated as a coordinate, just like position.

[6] For simplicity we treat each particle as a point so that we need not worry about its internal structure.

Hamiltonian for the overall system. The system may be very complicated, with particles of different masses moving with different velocities in all directions. The particles may collide and even attract or repel one another from a distance.

The beauty of the method is that it by-passes the effects of all the other particles, as we consider the six coordinates of every single particle in turn, and calculate how they relate to the overall Hamiltonian. Provided we have that Hamiltonian, we can obtain the equation of motion for each particle.

What all this means, at least in principle, is that from the Hamiltonian we can forecast the future and reconstruct the past! But what is true in principle is not always true in real life. We need to know two things: the form of the Hamiltonian and the precise initial position and momentum of every single particle. It is a scientific crystal ball, but like a 'real' crystal ball, is limited by the information it has at the beginning.

Hamilton's method does not add anything new to the laws of Newton and does not necessarily make it easier to solve practical problems. However, it does provide a deeper understanding of physical laws.

Conservation of energy

The mathematical representation on the blackboard below, which comes directly from Hamilton's theory, is an example of how his method provides a deeper understanding. The equation refers to how the total energy of an enclosed system, such as our system of particles, varies with time. The 'language' of the equation on the blackboard is the language of calculus where d/dt is an 'operator' which represents the rate of change with time of whatever quantity follows it, for whatever reason.

There is another operator which also represents rate of change with time, but in a more restricted sense; this is written as $\partial/\partial t$ and represents the rate of change specifically due to the passage of time, and not for any other reason.

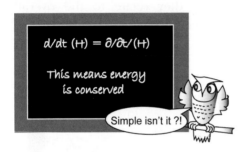

The left-hand side of the equation represents the rate of change of the total energy H, due to all causes. This *should be* equal to a series of terms on the right-hand side, representing changes of H due to specific changes in position and momentum of each of the individual particles. Hamilton's equations give us these terms but, when we put them in, we find that they all cancel and we are left with just the term representing the rate of change specifically due to passage of time.

The interpretation of this equation, which looks almost insignificant, is far from trivial. It means that although the individual particles may bounce around and interact, exchanging energy among themselves and continuously changing their positions and momenta, the total energy of the system is unaffected. That energy, H, can change only due to a specific cause such as putting the whole system into a magnet or perhaps into a vibrating food mixer. When there is no such intervention from outside, the total energy remains constant.

Last, but not least, is the feature which gave the Hamiltonian a central role in the quantum adventure. Hamilton's equations are based on the principle of least action, which turned out to apply in the world of atoms, just as in the household world. Some 50 years after Hamilton's death, his equations would be adapted seamlessly into quantum mechanics. The Hamiltonian would still be the crystal ball but, as we shall see, slightly clouded by the *uncertainty principle*.

William Rowan Hamilton

Hamilton was born in Dublin and was a child prodigy. Before he was three years old, William was sent to live with his uncle, James Hamilton, an Anglican priest who was in charge of a diocesan school in the town of Trim, about 50 km from Dublin. James was an accomplished linguist and an inspiring teacher. At the age of four, young William had made some progress in Hebrew; in the two succeeding years he acquired the elements of Greek and Latin and when he was 13, he was in different degrees acquainted with 13 languages, including Syriac, Persian, Arabic, Sanskrit, Hindustani, Malay, French, Italian, Spanish and German and this list may not even be complete.[7]

Impressive as was his list of linguistic accomplishments, Hamilton's fame was destined to come from mathematics. From childhood he had exceptional powers of mental arithmetic which on a certain occasion he had an opportunity to test in competition with another child prodigy, Zerah Colborn (1804–1839).

Colborn came to Dublin from America, as part of a grand tour of Europe demonstrating extraordinary feats of mental arithmetic involving large numbers which he multiplied and divided at great

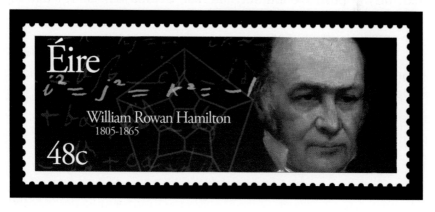

William Rowan Hamilton. Courtesy of An Post, Irish Post Office

---

[7] Dublin University Magazine 19, 1842.

speed. For example he is said to have calculated in his head that the number 4,294,967,297 is not prime,[8] but is divisible by 641 with 6,700,417 as the other divisor.

It appears that Colborn more often than not came out the victor in these contests, but the experience enhanced Hamilton's interest in mathematics and he immediately began to study the works of Euclid, Euler and Laplace. His study must have been pretty thorough as he found, and corrected, a significant error in Laplace's *Mécanique Céleste*!

In 1826 Hamilton graduated from Trinity College Dublin with the highest honours in both science and classics, having had the unique distinction of presenting a paper to the Royal Irish Academy, while still an undergraduate. The paper was entitled *Theory of Systems of Rays* and introduced what became known later as Hamilton's characteristic function for optics.

Even more unusual was his appointment as Professor of Astronomy at the University of Dublin, a position which carried with it the title of Royal Astronomer of Ireland, in 1827. Not surprisingly, there was an element of controversy that the position should be given to a 22-year-old, ahead of some other eminent and more experienced candidates. As it turned out, Hamilton was not especially fitted for the post, for although he had a profound knowledge of theoretical astronomy, he paid little attention to the regular work of the practical astronomer.

Hamilton took to writing poetry, but in a letter from the renowned William Wordsworth (1770–1850), with whom he had a long friendship, he received this tactful advice:

> *You send me showers of verses which I receive with much pleasure ... yet have*
> *we fears that this employment may seduce you from the path of science ...*
> *Again I do venture to submit to your consideration, whether the poetical*
> *parts of your nature would not find a field more favourable to their nature*

---

[8] This number is the fifth of a prime number sequence according to a formula proposed by Fermat in 1650. In 1732, Euler showed that Fermat was wrong, and that it can be factorized. It is possible that Colborn knew this and gave the answer from memory.

in the regions of prose, not because those regions are humbler, but because they may be gracefully and profitably trod, with footsteps less careful and in measures less elaborate.

The other great contribution Hamilton made to mathematical science was his discovery of quaternions. In 1844, his first paper entitled *On Quaternions; or On a new System of Imaginaries in Algebra* appeared in *The London, Edinburgh and Dublin Philosophical Magazine and Journal of Science.* Over the next six years there followed seventeen additional instalments, each instalment (including the last) ended with the words '*To be continued*'.

The algebra of quaternions is an extension of the algebra of complex numbers from two to four dimensions.

In Hamilton's own words:

Time is said to have only one dimension, and space to have three dimensions … The mathematical quaternion partakes of both these elements; in technical language it may be said to be "time plus space", or "space plus time": and in this sense it has, or at least involves a reference to, four dimensions. And how the One of Time, of Space the Three, Might in the Chain of Symbols girdled be.[9]

The explanation appears like an element of Einstein's theory of relativity but this is coincidental. Hamilton does not pretend to make a physical statement, but simply describes a mathematical technique. Quaternions lay dormant for a number of years; interest in them revived in the 20th century and they found application in both theoretical and applied mathematics. They are particularly effective in calculations involving a combination of space and time, such as three-dimensional graphics, computer simulation and satellite guidance systems.

---

[9] Robert Percival Graves. *Life of Sir William Rowan Hamilton.* Dublin University Press, Dublin. In three volumes; 1882, 1885 and 1889.

Chapter 3

The Pre-Quantum Atom —
A Temporary Solution

Going Outside the 'Household World'

We are familiar with aspects of a world of similar dimensions to ourselves; things which we can see and touch and be aware of. Let's call it the 'household world'. This world includes things that are a million times larger or smaller than we are, from ocean liners down to grains of sand.

Ancient civilizations were aware that there is a world outside the household domain. They even succeeded in getting remarkably accurate information about astronomical objects. Aristarchus (≈310–230 BC) measured the distance to the moon and estimated the distance to the sun, while Hipparchus of Rhodes (≈180–125 BC) produced a catalogue of about 850 stars. This quest was continued in the middle ages, when Copernicus developed a model of planetary motion and Newton applied his laws of motion to astronomical objects.

Exploring the world of the very small was much more difficult. A 'philosophical' approach was the only way open to the Greeks; the notion that atoms are the ultimate constituent of matter goes back to the 5th century BC, but there was little that could be said about them. The name comes from the Greek word *atomos*, meaning *indivisible*. Atoms were at the end of the line of things smaller and smaller, but the Greeks could not say where that end might be. The world of atoms was to remain hidden for a long time to come.

The Scientific Age

The first atomic theory of chemistry was presented in the books, *A New System of Chemical Philosophy,* by John Dalton (1766–1844) at the beginning of the 19th century. He was the first to treat atoms as physical entities rather than philosophical concepts. Dalton's atoms were hard, solid and indivisible and came in a variety of shapes and sizes; each chemical element had its own unique type of atom.

Messages from the Atoms

It was well known that light is emitted by hot objects such as the sun. Sunlight brings the energy to give us warmth and cause things to grow but perhaps it also brings messages from the inside of the sun, perhaps even from the ultimate atomic constituents? Maybe the colours are coded messages from the atoms.

Anders Jonas Ångström (1814–1874) was one of the founders of the science of spectroscopy. In a lecture to the Royal Swedish Academy of Sciences in 1853, he pointed out that the light emitted by gases at high temperature and vaporized metals comes at discrete wavelengths. These *spectral lines* are characteristic of the element, a unique fingerprint by which its presence can be recognized.

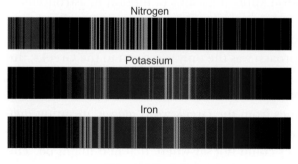

Characteristic spectra

He then demonstrated that when an electric spark jumps across a gap between two electrodes at high relative voltage, the light emitted consists of two superimposed spectra, one from the metal of the electrode and the other from the gas through which it passes.

Examples of the spectra of three elements are shown. The spectra were obtained with a *diffraction grating,* which gives much better resolution than the early pictures produced by sending the light through a prism. The spectra continue outside the visible range into the ultraviolet shorter wavelengths on the left and infrared on the right.

Conspicuous by Their Absence

Evidence of spectral lines had been discovered 50 years earlier, curiously not by their observation, but by their absence. In 1802, William Wollaston (1766–1828) noticed that the continuous, coloured band of light created when sunlight passed through a prism was punctuated by thin dark bands.

The bands were independently discovered by a German lens-maker, Joseph von Fraunhofer (1787–1826) in 1814. Fraunhofer measured the wavelengths of more than 500 of the bands, now known as *Fraunhofer lines.*

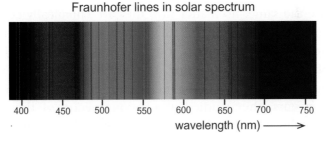

Fraunhofer lines in solar spectrum

wavelength (nm) ⟶

Fraunhofer lines — original sketch

Gustav Robert Kirchhoff (1824–1887) was the first to explain that the '*missing bits of light*' are evidence that something happens to the light before it reaches us. The dark lines in the sun's spectrum are caused by absorption of particular colours as the light passes through gases in the sun's atmosphere. Somehow the gas atoms capture light of those particular colours and let through all the other light through. The Fraunhofer lines mark the wavelengths where light has been absorbed.

In 1859, Kirchhoff and his colleague Robert Wilhelm Bunsen (1811–1899) began to study the spectra of metal salts vaporized in

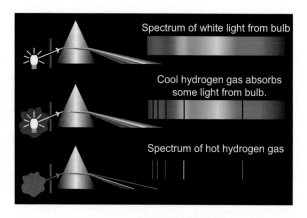

Emission and absorption spectra

the hot flame of a *Bunsen burner*. They found that the spectra of metals are the same, no matter how they are chemically combined. They were able to identify even minute traces of metals and soon discovered two new elements, caesium and rubidium. The technique of 'spectral fingerprinting' revolutionized chemical analysis. Realizing the importance of the method for the chemistry of the sun and other stars, they studied absorption spectra of the sun and identified the characteristic lines of several known chemical elements. Spectroscopic techniques remain our only means of studying the chemistry of the stars.

A Formula Without a Reason

In 1871, Ångström made accurate measurements of four lines in the hydrogen spectrum, at a time when there was a growing consensus that the wavelengths of spectral lines should somehow fit an empirical relationship.

Such a relationship emerged from a meeting between two friends, one of whom was Johann Jakob Balmer (1825–1898), a Swiss high school teacher with an aptitude for numerology. In the course of the conversation, he complained that he had '*run out of things to do*'.

His friend, who happened to be a physicist, replied: '*Well, you are interested in numbers, why don't you see what you can make of this set of numbers*'[1] and gave him the wavelengths measured by Ångström.

Balmer worked out that the wavelengths are fractions of a basic number H. In 1885, he published a surprisingly simple formula in which the wavelengths are expressed in terms of H and an integer n.

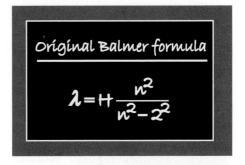

Original Balmer formula

$$\lambda = H\,\frac{n^2}{n^2 - 2^2}$$

The wavelength of the red line at 656 nm corresponds to $n = 3$ and the wavelengths of the remaining (blue) lines correspond to $n = 4$, 5, and 6. The formula applies equally well to other hydrogen lines not known to Balmer at the time and gives values which are at most $1/40,000$ of a wavelength different from the measured values.

Balmer had no physical understanding of how light is emitted, but his mathematical skill helped to solve the puzzle of atomic structure. In his first paper, Balmer wrote the prescient words: '*It appears to me that hydrogen … more than any other substance is destined to open new paths to the knowledge of the structure of matter and its properties …*'[2]

Glowing Gases

Discharge tubes were developed at the end of the 17th century. A high voltage across metal electrodes sealed into the ends of a

[1] *Archive for the History of Quantum Physics.* Interview with G.P. Thomson.
[2] Max Jammer. *The Conceptual Development of Quantum Mechanics.* McGraw-Hill, New York. 1966.

partly evacuated glass tube causes electrical breakdown of the residual gas in the tube. The discharge takes the form of a spark (similar to lightning) at high pressures and a diffuse glow at low pressures.

In the mid 19th century, Heinrich Geissler (1814–1879) developed a mercury vapour pump to evacuate discharge tubes down to a pressure of 1/10,000 of an atmosphere. When a high voltage was applied across the electrodes, a bright luminous glow appeared. The tubes were sold to schools and universities for demonstration purposes.

Geissler began to experiment, filling his tubes with gases, vapours and liquids; a talented glassblower, he made tubes in exotic and beautiful shapes. The tubes captured the imagination of the public and many were sold as decorative pieces. The *Geissler tube* is the ancestor of the neon sign.

Mysterious Rays

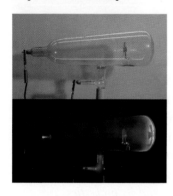

Crookes tube. Courtesy of D-Kuru/Wikimedia Commons

William Crookes (1832–1919) was one of the first to investigate what happens when most of the air is pumped out of a discharge tube. He found that when the pressure in the tube is about one millionth of an atmosphere, almost no gas remains but the tube still glows. Crookes established that the glow was produced by 'rays' coming from the *cathode* (negative electrode) and they became known as *cathode rays*. He showed that cathode rays travel in straight lines and cause a fluorescent glow on the glass behind the *anode* (positive electrode). The photograph shows the best known Crookes tubes in which the anode is in the shape of a Maltese cross. The shadow of the cross on the end of the tube shows that cathode rays cannot penetrate a metal; the size and position of the shadow shows they travel in a straight line.

Something Smaller Than an Atom — The Discovery of the Electron

Cathode rays remained something of a mystery until 1897, when John Joseph (J.J.) Thomson (1856–1940) established that they were streams of 'corpuscles', which carried negative electric charge of the same size as the positive charge on a hydrogen ion. Thomson used electric and magnetic fields to manipulate the particles and made the surprising discovery that: '*... the mass of the corpuscle is only about 1/1,700 times the mass of the hydrogen. Thus the atom is not the ultimate limit to the subdivision of matter.*'

The corpuscle of electricity was the electron — the first 'sub-atomic' particle to be discovered. For his discovery Thomson was awarded the Nobel Prize in 1906. The word *electron* was coined by George Johnstone Stoney (1826–1911). It derives from the Greek word κίτρινο ηλέκτρου, meaning amber. Amber was known to attract light objects such as bits of paper after being rubbed with a dry cloth (an example of static electricity).

Where Do These Electrons Come From?

The electrons appeared as if from nowhere; they must have been liberated from atoms in the cathode of the discharge tube, which begs the question '*How do electrons fit into atoms in the first place?*'

Atoms are stable and generally uncharged so if the negatively charged electrons are inside the atom, they must somehow be balanced by an equivalent amount of positive charge.

Thomson's 'Plum Pudding' Model of the Atom

In 1904, J.J. Thomson came up with the following proposal for accommodating the electrons in an atom: '*the atoms of the elements consist of a number of negatively electrified corpuscles enclosed in a sphere of uniform positive electrification …'*

Thomson's atom

Thomson proposed that an atom contains enough positive charge to cancel the total negative charge of the electrons. The positive charge (and almost all the mass) is spread uniformly over a sphere containing the electrons. Such an atom would certainly be stable. (Thomson's model is often called the *plum pudding* model because the electrons are 'strewn around' like raisins[3] in a plum pudding.)

He did not develop the model further as at that time there was no experimental evidence and any refinements of the model would have amounted to pure speculation. James Clerk Maxwell (1831–1879) had made the theoretical prediction that oscillating electric charges emit energy in the form of *electromagnetic waves* which propagate through empty space at a speed equal to the speed of light. If this was the mechanism by which atoms emit light then, according to the theory, the frequency of the radiation would be equal to the frequency of the oscillating source.

As regards atomic spectra, the model does not score well. Suppose that the electron in a hydrogen atom is disturbed from its natural place at the centre of the atom; it will oscillate — trapped inside the atom by the attractive force of the positive charge — like a ball bearing in a bowl. Light may well be emitted as the electron oscillates but there is just no mechanism for producing a characteristic spectrum of different coloured lines. That aspect of the model was not pursued at the time.

Rutherford Investigates

Ernest Rutherford (1871–1937) received the Nobel Prize in 1908 *'for his investigations into the disintegration of the elements, and the chemistry of radioactive substances'*. He continued to study the behaviour of

---

[3] Dried plums used in making puddings and cakes.

alpha particles, the positively charged helium ions emitted by the disintegration of heavy radioactive elements.

At Rutherford's laboratory in Manchester, he and Hans Geiger (1882–1945) were bombarding thin gold foils with alpha particles from the radioactive element radium. A fluorescent screen, placed round the apparatus, emitted an individual flash of light each time an alpha particle landed. Rutherford and Hans Geiger, a postdoctoral fellow, would sit in the dark counting the flashes of light produced by alpha particles as they hit the screen. Geiger later developed electronic equipment — the *Geiger counter* — to perform the tedious task of counting alpha particles.

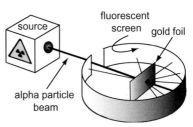

Geiger and Marsden's experiment

If Thomson's model were correct, the positively charged alpha particles would barge their way through the thin foil, being only slightly deflected by localized excesses of either positive or negative charge. They should emerge from the foil more or less undeflected.

An Unexpected Observation

In 1909, Ernest Marsden (1889–1970), an undergraduate student, was assigned what turned out to be the most exciting undergraduate project of all time; he was co-opted to help Geiger to search for particles scattered through large angles. It must not have seemed very exciting at first, as large angle scatters were few and far between. Then, to everyone's surprise, Geiger and Marsden found that a small fraction of alpha particles, maybe 1 in 8,000, were bounced right back in almost the direction they had come. Rutherford recalled the event in a lecture some time later:

> *It was quite the most incredible event that ever happened to me in my life. It was almost as incredible as if you fired a 15-inch shell at a piece of tissue paper and it came back and hit you.*

Rutherford Interprets the Results

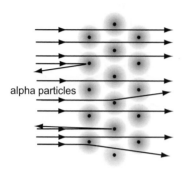

Rutherford scattering

In 1911, Rutherford interpreted Geiger and Marsden's results in terms of a *nuclear* atom with all the positive charge and nearly all the mass concentrated in a very tiny nucleus. He estimated the central charge of a gold atom as about 100 e (50 times the charge of an alpha particle). The massive and highly charged gold nuclei strongly repel alpha particles which come very close to them, explaining why a few alpha particles are scattered through such large angles:

> *It seems reasonable to suppose that the deflexion* (sic) *through a large angle is due to a single atomic encounter ... A simple calculation shows that the atom must be a seat of an intense electric field in order to produce such a large deflexion* (sic) *at a single encounter.*

How Small is Small?

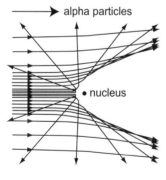

Alpha particle trajectories

Rutherford then estimated the maximum size of the nucleus by applying classical physics to the interactions between nuclei and alpha particles. He calculated the angular distribution to be expected on using the inverse square law of electrostatic repulsion discovered by Charles Coulomb (1736–1806).

The 'closest encounter' occurs when the alpha particle is heading directly towards the centre of the nucleus and is effectively brought to a stop before bouncing back in the direction from which it came. That specific particle has used up all its kinetic energy against the repulsive electric force of the nucleus. Rutherford knew the initial kinetic energy of the incoming particle as well as the electric charge of both the alpha particle and the nucleus. From this, he calculated the distance between their centres when the alpha particle has been brought to a stop. The sum of the radii of the nucleus and alpha particle had to be smaller than that.

The maximum radius of the nucleus turned out to be about 1.5×10^{-14} m, compared with an atomic radius of about 1.5×10^{-10} m. This means that the nucleus is at least 10,000 times smaller than an atom — atoms are

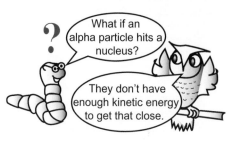

mostly empty space. No wonder almost all the alphas went straight through the foil!

So far, Rutherford's model scores well — he was able to estimate both the charge and the maximum size of a gold nucleus. His estimate of the diameter was about three times larger than the current value, a good approximation at the time. Rutherford's great contribution lay in showing that the Thomson model of the atom cannot possibly explain the large-angle scatterings, whereas the nuclear model can.

The Planetary Model

If the nucleus is tiny compared with the atom, then it follows that the electrons must be somehow distributed in the empty space around the nucleus.

Rutherford proposed a 'planetary' model in which electrons move in circular orbits about a nucleus. Each electron revolves at a speed such that the electrostatic force of attraction is just sufficient to keep it in orbit. The dynamics of the model are familiar enough because they are similar to Newton's model of planetary motion.

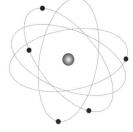

Planetary model

Relative Sizes

Wembley stadium

To get a better idea of the scale of the atom, let us imagine that the nucleus is represented by a marble of diameter 2 cm, placed at the centre of the football field. In such a scale model, the atomic electrons will move in orbits which are outside the stands!

A Question of Stability

The problem with Rutherford's model is that the planetary atom, unlike the solar system, is just not stable. It was well known from Maxwell's equations that electrons lose energy by emitting electromagnetic radiation every time they change their direction of motion; orbiting electrons change direction constantly and would be expected to radiate energy continuously. In this way, they would lose energy and spiral towards the nucleus, pulled in by electrostatic attraction; the atom should collapse.

Rutherford did not address this open question of the stability of the nuclear atom. He side-stepped the problem and produced an uncharacteristically vague statement:

> *The question of the stability of the atom proposed need not be considered at this stage, for this will obviously depend upon the minute structure of the atom, and on the motion of the constituent charged parts.*

A Historical Note

Ernest Rutherford graduated from the University of New Zealand in Wellington in 1894. He was awarded an 1851 Exhibition Science

Scholarship and went to the Cavendish Laboratory in Cambridge to continue his studies under J.J. Thomson. In 1898, Rutherford became Professor of Physics at McGill University in Montreal. He identified two of the three main components of radiation from the radioactive decay of atoms and named them alpha rays

Lord Rutherford. Courtesy of New Zealand Post

and beta rays. He also showed that alpha particles[4] were positively charged helium atoms.

At McGill, Rutherford worked with Frederick Soddy (1877–1956), studying the decay of radium ('thorium X') into an inert gas, which they figured was most probably one of the noble gases discovered between 1894 and 1898. Soddy later recollected part of a conversation between them:

Soddy: *Rutherford, this is transmutation: the thorium is disintegrating and transmuting itself into an argon gas ...*

Rutherford: *... don't call it transmutation ... they'll have our heads off as alchemists.*

In 1908 Rutherford was awarded the Nobel Prize in chemistry. In his after dinner speech at the Nobel banquet, Rutherford remarked:

I have dealt with many different transformations with various periods of time, but the quickest that I have met was my own transformation in one moment from a physicist to a chemist.

Rutherford returned to England that same year to become Professor of Physics at Manchester University. There he supervised the famous alpha particle scattering experiments which led to the *Rutherford model* of atomic structure.

---

[4] The terms 'particle' and 'ray' were used interchangeably at this time.

Rutherford later used alpha particles to bombard nitrogen gas, producing oxygen atoms and protons — the first example of transmutation of nitrogen into oxygen and the first artificially induced nuclear reaction. He went to the Cavendish Laboratory in Cambridge in 1920, to become Professor of Experimental Physics and remained active in research to the end of his life.

Lord Rutherford. Courtesy of New Zealand Post

Rutherford received many tributes both during and at the end of his life:

> *With the passing away of Lord Rutherford, the life of one of the greatest men who ever worked in science has come to an end … He left science in quite a different state from that in which he found it … He will be missed more, perhaps, than any scientific worker has ever been missed before.*

Niels Bohr. October 1937

Chapter 4

Max Planck — The Birth of the Quantum Adventure

On 21 October 1874, a 16-year-old student named Max Planck (1858–1947) entered the University of Munich. He chose to study physics, but was warned by his professor, Philipp von Jolly (1809–1884) that this branch of *Natural Philosophy* offered no real prospects. The laws of physics were fully understood. Nothing new remained to be discovered.

Von Jolly had some justification for his opinion. Most aspects of the laws of nature appeared to be understood, or at the very least, categorized. They presented a beautiful picture, logically interconnected like a gigantic jigsaw, put together by the natural philosophers of the preceding centuries. The fundamental laws of motion, which Isaac Newton had formulated almost 200 years earlier, were concise and simple and held with amazing accuracy. They had been applied to objects big and small and, when combined with Newton's law of gravitation, explained even the motion of heavenly bodies such as the Earth and planets around the sun. As an added bonus, Newton's mechanics had subsequently been shown to relate to a basic economy of nature and expressed in beautiful mathematics by Lagrange and Hamilton.

Thermodynamics

Thermodynamics, the branch of physics that deals with relationships between heat, temperature, work and energy, seemed also to be completely understood. All thermodynamic processes had been embraced by a single universal theory which comes from the logical development of just three fundamental laws.

47

Electromagnetism

Ten years earlier, in 1864, the Scottish physicist, James Clerk Maxwell (1831–1879), had synthesized the laws governing electricity and magnetism and shown that vibrating electric charges send out electromagnetic waves, which propagate with the speed of light. His beautiful theory was summarized in the four equations which bear his name. These equations led to the prediction of electromagnetic radiation, of which visible light forms a small part. The mechanism of the creation of *light* was understood; apparently there was little left to discover.

Heads in the Sand?

Was critical evidence being ignored?

Von Jolly's complacency was shared by a number of natural philosophers at that time and into the early years of the 20th century. There were others, however, who were not prepared to 'close the book' on developments in physical science. They realized that the gigantic jigsaw of natural phenomena was incomplete; some pieces did not fit together properly; others did not fit at all.

Despite von Jolly's warning, Planck decided to study physics at the University of Munich. He wrote his doctoral thesis at the very early age of 21 and soon afterwards obtained an academic post at the University of Kiel.

An Incomplete Jigsaw

In hindsight, we can identify the main unsolved problems that were being ignored. These gaps in understanding, in diverse areas of

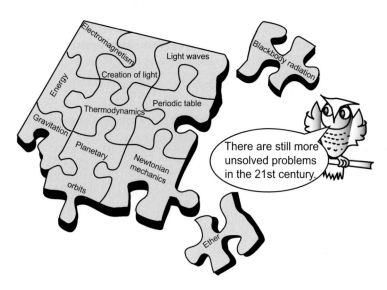

The jigsaw of natural philosophy at the start of the 20th century

natural philosophy may have seemed unimportant at the time, but were soon to prove of critical significance.

Two Missing Pieces

The ether

There was an implicit assumption that all waves require a medium in order to propagate. Sound, for example, travels through air and through solids and liquids but does not propagate through a vacuum. The notion of a wave without a medium was considered to be absurd.

Light waves reach us from the sun, stars and distant galaxies across billions of miles of interstellar space; a medium had to be invented to fill that space. It was therefore believed

Hard to imagine a wave without a medium

that the universe was filled with a substance called 'ether', with no known properties other than the ability to carry light.

At what is now the Case Western Reserve University in Cleveland, Ohio, Albert Michelson (1852–1931) and Edward Morley (1838–1923) were trying to find evidence of this ether. They carried out experiments to detect the 'ether wind' which would blow as the Earth hurtles in its orbit around the sun.

The method was very clever; it was based on the premise that light should go faster with a following ether wind and more slowly when struggling against it. The difference in speed could be measured with great accuracy by studying the interference[1] between two light waves, which had been sent on return journeys in different directions and then recombined. If there were an ether wind, its effect would depend on the orientation of the apparatus.

Michelson had designed and built the apparatus in Berlin in 1881, while on study leave from the US Navy. He resigned from the navy in 1883 and returned to the USA to become a professor at Case Western. There, he and Morley made substantial modifications to improve the sensitivity of the original apparatus and repeated the experiment in 1887.

'The greatest failed experiment in history'

The *Michelson–Morley* interferometer was sensitive enough to detect differences as small as 15 km/hr (jogging speed) in the speed of the two light waves. Every precaution was taken to ensure accuracy. The experiment was repeated at different times of the day and in different seasons of the year. Still, in 1887 as in 1881, there was no evidence of an ether wind. Could it be that Galileo, Newton and Copernicus were all wrong and the Earth was *not* in orbit around the sun, but was stationary at the centre of the universe?

The ether was never found. There is no place for it in the jigsaw. The fact that no one could find it was not a failure, but a great success. It led to one of the major scientific developments of the 20th

---

[1] The phenomenon of interference is described in Chapter 5.

century, when Albert Einstein wiped the slate clean of preconceived prejudices by assuming that there is no ether, and conceived the theory of special relativity. This theory was one of the great 'discoveries of the mind' and led to a new understanding of space and time; it is very relevant to the world of subnuclear physics where elementary particles travel at speeds close to the speed of light and matter is created out of energy. From relativity came predictions regarding the equivalence of matter and energy, with verifiable practical consequences.

The fact that the 'ether' piece of the jigsaw did not exist meant that there was room to fit a number of hitherto unknown pieces, which would be added later by Albert Einstein.

The world inside the atom

Another area where knowledge was incomplete was the domain of atoms and molecules. The existence of atoms was generally accepted. As far back as 1811, Amedeo Avogadro (1776–1856) had written about atoms and by 1871, Dmitri Mendeleev (1834–1907) had compiled the first draft of his periodic table of the elements. Atoms were considered as point objects with no inner structure and it was not clear what distinguished atoms of different elements.

There was no information on the laws of this 'atomic' world. Were individual atoms really simple point objects moving according to Newton's laws, or was there a structure, a hidden world, inside the atom? Such questions were not easy, perhaps even impossible, to answer; best to ignore them like the ostrich with its head in the sand.

As the 19th century drew to a close, the subatomic world remained largely unknown and undetected.

Light from Hot Coal

Then, there was another puzzle. What makes hot objects glow in the dark? This, at first, did not seem like a matter of major concern, but perhaps the answer would contain some vital clue to the world of

Coal at room temperature

Hot coals

*Black surface — an open window
to the inside*

atoms? Could it be that light was a messenger, bearing information about the basic structure of matter? This possibility was recognized towards the end of the 19th century. Nobody could have guessed that the light carried revolutionary news about a basic law of nature.

At room temperature a piece of coal looks black because it absorbs all the light which falls on it; no light is reflected. As its temperature rises the coal begins to glow, at first dimly and then more brightly, with rapidly increasing intensity. At the same time, the coal changes colour, going from deep red to orange, to blue and finally to 'white hot'.

In 1859, Kirchhoff proposed that surfaces which emit light most efficiently also absorb light energy completely. This follows from basic requirements of thermodynamic equilibrium; if there were more energy absorbed than emitted, an imbalance would build-up very quickly.

Blackbody Radiation — A Window to the Atom

A black surface acts like an open window, absorbing all radiation that falls on it and reflecting none.

From the outside, the window appears black, unless there is light coming from the inside. At the same time, if there *is* any light inside the house, it will come out without diminution and will bring information from inside. Kirchhoff introduced the name *blackbody radiation* to describe the light from such surfaces.

Kirchhoff's hypothesis may appear obvious and logical. If coal were a good absorber but a poor emitter, it would act as an energy trap and become red hot very quickly if left in the sun. However, he went on to say what is not so obvious, but very

important, that the relative amount of light emitted at different wavelengths depends on nothing else other than the temperature.

More precisely: the intensity distribution of blackbody radiation over the spectrum of wavelengths is universal, depending only on the temperature and not on the size, shape or material of the object.

What We See Through the Window

The figure illustrates how blackbody radiation varies with wavelength.

The intensity of the radiation is plotted as a function of wavelength for three different temperatures, 2,500 K, 3,000 K and 3,500 K.[2] The area under each curve represents the total intensity of radiation emitted at all wavelengths; we can

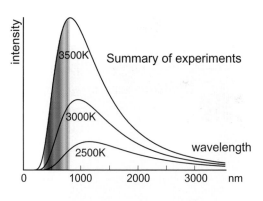

The spectrum of light from a hot body

see that it increases rapidly with temperature. In 1879, Josef Stefan (1835–1893) of the University of Vienna, published a quantitative

---

[2] Temperatures are usually quoted with reference to the 'absolute' *Kelvin* scale of temperature (K), rather than the more familiar *Celsius* scale. Zero Kelvin is defined as 'the theoretical absence of all thermal energy' and is equivalent to −273°C.

estimate of this increase; he found that the total intensity of radiation depends on the fourth power of the absolute temperature.

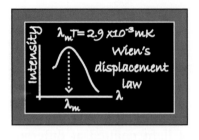

In Berlin, Wilhelm Wien (1864–1928) analysed the experimental data further. He established that as the temperature rises, the highest point of the intensity graph moves to the left — towards lower wavelengths — in a well-defined way.

Wien's displacement law states that the product of λ_m, the wavelength at which the radiation energy is greatest, and T, the temperature of the radiating surface is the same for all black surfaces. As the temperature increases, this dominant wavelength decreases and the colour of the visible spectrum changes from red to white to blue.

Measuring Temperature from a Distance

Star Cluster NGC 265. Courtesy of ESA, NASA and E. Olszewski (University of Arizona)

A practical application of Wien's law is the measurement of the temperature of hot glowing objects from the colour of the radiation they emit. This can be achieved by matching the brightness of a tungsten lamp filament with the brightness of the hot object. *Optical pyrometers,* which use this technique, are used to measure temperatures of for example furnaces or molten steel. The advantage of tungsten is that its melting point is approximately 3,700 K, much higher than that of iron (1,800 K) or copper (1,350 K). To measure even higher temperatures, the brightness of the source is reduced by using a filter.

Distance is no object in optical pyrometry; the technique can be used to measure the surface temperature of the sun and of other stars within and outside our galaxy.

The sun has a surface temperature of about 5,800 K. This means that the maximum of the radiation spectrum is right in the middle of the optical region; not really surprising from the point of view of evolution of life on this planet!

And I thought all stars looked the same.

The picture above shows some of the stars in the Sagittarius cloud. The relatively faint orange and red stars look much the same as our sun would appear if viewed from another part of our galaxy. The bright red stars are cool *red giants*, once similar to the sun but now in a more advanced stage of evolution. The blue and greenish stars are hotter, many of them relatively young and massive. The Sagittarius cloud of stars lies towards the centre of our galaxy. It transpires that the surface temperatures of stars cover a wide range, up to more than 200,000 K.

Making a Theoretical Model — 'Cavity' Radiation

How does one even begin to calculate radiation from a hot surface theoretically? The classical way of simulating the process is to use what looks like an unlikely model of an enclosed cavity, such as an oven, with its inside walls at a high temperature. According to Maxwell's theory, such a cavity is a perfect model; electric charges in its walls vibrate (almost as if they were attached to invisible springs) and emit electromagnetic radiation at various frequencies. The radiation bounces around the cavity, sometimes reflected and sometimes absorbed by the walls until a state of equilibrium is reached, when the rate at which energy is emitted by the walls is equal to the rate at which it is absorbed.

The radiation inside the cavity can then be defined as *equilibrium blackbody radiation* corresponding to a particular oven temperature; it does not depend on the size or shape of the cavity, nor on the material of its walls. A small opening in the wall of the cavity acts as a perfect emitter, a window through which a sample of the radiation leaves the cavity. It is also a perfect absorber; all the light falling on the opening is absorbed and then bounces around the cavity.

Wien's Spectral Distribution Law — One Step Further by Thermodynamics

Using the principles of thermodynamics, Wien was able to take another step forward by predicting an interesting feature of black-body radiation. In 1896, he showed that the intensity of the radiation, divided by T^5, the fifth power of the temperature, depends in some way on λT, the *product* of wavelength and temperature, and nothing else. At this stage, Wien did not know the exact relationship; only that it existed.

Wien realized that he had found something special, a way of describing how the intensity of blackbody radiation varies, in the form of a single graph, a *universal curve*, which is valid for *all* temperatures.

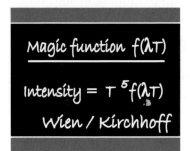

The maximum point on the graph occurs at $\lambda T = 2.9 \times 10^{-3}\,\text{mK}$, nicely confirming the experimental observations that Wien himself had made in 1879. It is a general statement of his *displacement law* and it applies to all wavelengths and temperatures, not just to the point at which the radiation intensity is greatest.

Wien had found the 'magic' function which completely describes the blackbody spectrum. Kirchhoff had suspected that it must exist and had essentially challenged theoreticians to find it. Using thermodynamic arguments, Wien deduced how to plot the function in the form of a single universal graph.

An Experimental Cavity Radiator

A number of German experimentalists immediately set out to measure the energy density of the radiation emitted by an experimental model of a cavity radiator. At the Physikalisch-Technische Reichsanstalt (PTR) in Charlottenburg (Berlin), Otto Lummer (1860–1925) and Ferdinand Kurlbaum (1857–1927) constructed a metal box with electrically heated walls and a small hole at the side. They were able to study the outgoing radiation and to make remarkably accurate measurements of the heat energy emitted from the apparatus.

Lummer and Kurlbaum's apparatus

Simultaneously, Heinrich Rubens (1865–1922) a guest scientist at the PTR in Charlottenburg and Friedrich Paschen (1865–1947) at the Technische Hochschule in Hanover carried out similar experiments. The radiation from the model cavity turned out to be indistinguishable from the radiation emitted by black surfaces, confirming the Stefan–Boltzmann law.

Measuring the Radiation

There is little point in building a cavity radiator unless one has an accurate method of measuring the small amount of radiation which emerges through the opening. The American astronomer Samuel Pierpont Langley (1834–1906) is credited with the invention of an apparatus to measure such small radiation intensity (the *bolometer*). He used it in 1878 to measure the intensity of starlight.

Langley is better known as one of the early pioneers of aviation. He began by building models of heavier-than-air flying machines powered by miniature steam engines, built by Charles Manly. In 1887, he succeeded in flying one such model for a distance of more than 1 km, ten times further than any previous flight. Two piloted attempts in 1903 were not so

Langley and Manly (left). Courtesy of NASA

successful. Both machines crashed on take-off, fortunately with no major injuries. Manly was recovered unhurt from the river on both occasions.

In 1890, Langley founded the Smithsonian Astrophysical Observatory. Named in his honour are the Langley NASA Research Centre, the Langley Air Force Base, the Langley unit of solar radiation and Mount Langley in Sierra Nevada.

Experiments Confirm There is a Magic Function

In 1899, Lummer and Ernst Pringsheim (1859–1917) published some remarkable experimental data in the proceedings of the German Physical Society. They had made measurements at three different temperatures: 1,259 K, 1,449 K and 1,646 K and had found that the values of the magic function do, in fact, lie on a single curve.

What an odd situation! On the one hand, thermodynamics was saying that the magic function exists, but did not say what it was. On

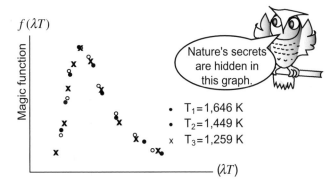

Lummer and Pringsheim measurements 1899.

the other hand, there was experimental data which clustered around a single curve, but no one knew why that curve was the shape that it was!

Where to Go from Here?

Was the light telling us something about the atoms vibrating inside hot glowing matter? Was it possible to crack the code of the blackbody spectrum and discover the mechanism by which light is produced?

Wien noticed a striking similarity between the shape of the blackbody curve and the shape of another well-known, but totally unrelated, curve showing the distribution of speeds of the molecules of a low pressure gas.

The gas molecules are thermally agitated and move around randomly occasionally colliding and transferring energy from one to another.

Wien's formula

$$f(\lambda T) = \frac{a}{(\lambda T)^5 e^{b/\lambda T}}$$

As Wien described in his lecture, he guessed that blackbody radiation is emitted from electrons moving in the same sort of way as the gas molecules which: '*emit radiation of wavelength dictated by velocity*'.

In 1896, Wien published an ad hoc formula for the mysterious function. His formula was quite successful, particularly for short wavelengths. Wien had no convincing physical explanation for the form of the function, and further experiments at long wavelengths gave results which did not agree with his formula.

Wien made an educated guess at the mathematical form of $f(\lambda T)$ and introduced two constants a and b, which he could adjust to make the function agree with the experimental data.

The Rayleigh–Jeans Law

In the meantime, in Cambridge, an English physicist, John William Strutt, Lord Rayleigh (1842–1919), and the young mathematician James Jeans (1877–1946) were trying a more direct physical approach to cavity radiation. They proposed that radiation emitted from the walls of a cavity is reflected back and forth inside the cavity and a system of standing waves is eventually set up, rather like the sound waves in the resonant cavity of a musical instrument. As the wavelength becomes shorter, the number of possible *modes of vibration* (standing waves) increases.

Rayleigh and Jeans calculated the density per unit wavelength of modes, at all wavelengths, and found it to be independent of the shape of the cavity. Applying the well-established thermodynamic principle of the equipartition of energy, they too obtained an equation for the 'magic' function.

Rayleigh and Jeans' expression looks completely different from Wien's magic function; nevertheless it fits the experimental data extremely well for longer wavelengths. By contrast, the curve shoots off to

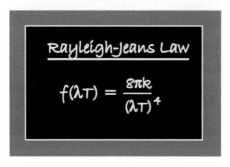

Rayleigh–Jeans Law

$$f(\lambda T) = \frac{8\pi k}{(\lambda T)^4}$$

infinity as the wavelength becomes very small; this implies that a rapidly increasing amount of energy is radiated as the wavelength decreases. The physical consequence would be that X-rays and γ-rays (gamma rays) of great intensity would be produced by burning coal. Something was very wrong with the formula. It predicts an *ultraviolet catastrophe.*

It is an indication of the lack of urgency attached to the subject, that cavity radiation was by no means the only interest of either Rayleigh or Jeans. Lord Rayleigh is arguably better known for his work in optics and published papers on such diverse topics as photography, colour vision, sound, electromagnetism and hydrodynamics. He received the Nobel Prize in physics in 1904 for '*the investigation of the densities of the most important gases and for his discovery of argon in connection with these studies*'. Rayleigh craters on the moon and on Mars are named in his honour.

Sir James Hopwood Jeans' main interests were in astronomy and stellar evolution. He wrote books on the theory of gases, theoretical mechanics and electricity and magnetism and the relation of science to philosophy. He also has craters named after him, on the moon and on Mars.

Two Theories Each 'Half Right' and 'Half Wrong'

There were now two theories, each of them giving good predictions but only for opposite halves of the spectrum. Wien's formula might appear to give a better all round fit, but it contains two arbitrary constants which give some freedom of adjustment. The Rayleigh–Jeans law had to be treated more seriously but its divergence from reality was catastrophic at short wavelengths. The sketch shows the theoretical curves compared to the 1899 experimental data.

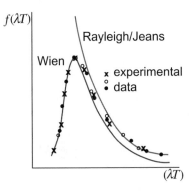

Which one to choose?

Planck's 'Inspired Guess'

Planck, who had been appointed Professor of Theoretical Physics at the University of Berlin in 1889, became fascinated by the interaction between light and matter. There were so many questions to answer. Why does the colour of hot objects change with temperature? Is light from hot surfaces really bringing messages from the atoms inside? Are there clues in this light, if only we could decipher them? Planck tried to find a physical basis for Wien's formula, but without success.

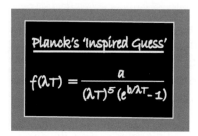

Planck's 'Inspired Guess'

$$f(\lambda T) = \frac{a}{(\lambda T)^5 (e^{b/\lambda T} - 1)}$$

In the summer of 1900, Kurlbaum and Rubens made very accurate new measurements of the spectrum of heat radiation. When Planck heard about these results he felt an even greater urgency to pursue the problem. Leaving physical reasoning aside, at least for the moment, he turned to some mathematical manoeuvring. Quite quickly these efforts enabled him to construct a formula which reduces to Wien's formula at short wavelengths and, at long wavelengths, to the Rayleigh–Jeans law. It involved a minimal adjustment to the denominator in Wien's formula, and created a mathematical link between the two models. It was not a solution to the problem but created a mathematical link between them.

A short time afterwards, Planck and Rubens met for tea in Planck's home and compared Rubens' latest results with a new formula suggested by Planck. The comparison showed complete agreement throughout the entire range of wavelengths. Planck's excitement was re-kindled; as he wrote later:

> *without the intervention of Rubens the foundation of quantum theory would have perhaps taken place in a totally different manner, and perhaps even not at all in Germany.*[3]

---

[3] Jagdish Mehra and Helmut Rechenberg (eds). *The Historical Development of Quantum Theory.* Volume 1. Springer, Berlin. 2001.

Planck presented his formula as a 'comment' entitled *On an Improvement of Wien's Radiation Law* at a

meeting of the German Physical Society on 19 October 1900. He did not claim that it was based on any physical reasoning or that it added anything to the understanding of the problem, and later called it 'an inspired guess'. It was, nevertheless, a link which unified what had been done before.

The majority of physicists paid scant attention to Planck's 'comment'. After all, it was just a simple mathematical manipulation, with no physical content. Planck however had the belief that he had made a step in the right direction.

Planck Looks for Something Deeper

Planck's instinct told him that his empirical formula could not be simply 'a lucky guess'. He worked intensely, searching for a code concealed in the mathematical expression, which would lead him back to nature's secret, hidden in the messages which light brings from hot surfaces. At first he tried classical arguments, based on physical principles which had stood steadfast over the previous centuries:

> *I was filled with what would be thought to-day naively charming and agreeable expectations that the laws of classical electrodynamics would ... enable us to grasp the most significant part of the process ... the problem towered with fearsome height even steeper before me ...*[4]

The Rayleigh–Jeans function was derived from sound physical reasoning and was the obvious place for Planck to start. Their hypothesis that radiation originates in the oscillations of electric charges in the walls of the cavity was based on Maxwell's beautiful and already well-established theory of electromagnetism, and Planck was convinced that it was correct. Why then was their calculation of

---

[4] This and subsequent quotations, unless otherwise indicated, are taken from Planck's Nobel Prize acceptance speech (1918).

radiation intensity too high at larger frequencies? Was there something they were not taking into account? Something which was preventing large numbers of higher frequency oscillations? Perhaps there was some law of nature which limits the number of oscillators vibrating at high frequencies.

Planck had exhausted all classical solutions and decided that an '*act of desperation*'[5] was necessary:

> *A theoretical interpretation had to be found at any price, no matter how high that might be. After some weeks of the most strenuous work of my life, light came into the darkness, and a new undreamed-of perspective opened up before me.*

The Moment of Truth — Planck's Quantum Hypothesis

The 'perspective' was a simple but radical hypothesis: nature prevents an ultraviolet catastrophe by placing restrictions on the allowed energies of oscillators. The higher the frequency, the greater the restriction, because the gap between one allowed energy and the next increases.

Planck presented his new perspective to the German Physical Society on 14 December 1900, less than two months after his previous communication:

> *An oscillator can only have energy consisting of an integral number of quantum units. The value of each unit depends on the frequency of the oscillator and is given by a simple relationship, $E = hf$. h is a universal constant, now called Planck's constant, and f is the frequency of the oscillator.*

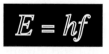

Atom of energy.

It was the beginning of the *Quantum Adventure*. Although not yet ready to speculate on the consequences of the

---

[5] Unpublished letter addressed to R.W. Wood. Centre for History and Philosophy of Physics, American Institute of Physics, New York. 7 October 1931.

hypothesis, Planck realized immediately that it entailed a fundamental change in the philosophy of physics. The concept of atomicity of matter and the more recent atomicity of charge were
extended by a new hypothesis of indivisible and discrete units of
energy. The initial hypothesis applied specifically to the energy of
an oscillator.

An Inspired Guess Becomes a Physical Theory

Planck later described how he got a very pleasant surprise when
he introduced the quantum condition into the Rayleigh–Jeans
formula. It transformed into exactly the same form as his empirical
formula. This time it was not an inspired guess, but had come from
a physical argument.

Rayleigh and Jeans had based their derivation on the classical
energy distribution of oscillators in the walls of the cavity which is
continuous and decreases exponentially with energy. Planck's quantum condition allows only energies with magnitudes which are multiples of the quantum hf. An oscillator can gain or lose energy only
in discrete units of size hf, otherwise it is left in a forbidden energy
state. The effect of restricting the values of energy in this way
becomes more noticeable at high frequencies where the allowed
energies are further apart.

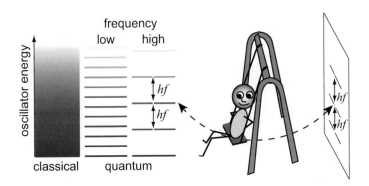

The electron swing is restricted by the allowed energies of a quantum oscillator

Planck's Constant

Planck adjusted the value of h in his formula until he got the best fit between his curve and the experimental data. He obtained a value of 6.55×10^{-34} Js, close to the modern value of 6.626069×10^{-34} Js.

Planck's constant is expressed in units of energy × time and has the same dimensions as action. It is called the *elementary quantum of action*.

How Do the Oscillators Lose Their Energy?

Maxwell's prediction that an oscillating charge emits energy by radiation was verified experimentally in 1887 by Heinrich Hertz (1857–1894). Since all electromagnetic waves carry energy, an electric oscillator loses energy as it emits electromagnetic radiation. Classically the energy loss is smooth and continuous and the oscillator energy gradually drops to lower values (levels).

It was not clear how the quantum condition would affect the process. If only certain energy levels were possible, losing energy would certainly not be smooth, but rather like tumbling down stairs. Electromagnetic energy would then be released in discrete bursts.

For a number of years Planck made no assertions as to the full implications of his hypothesis. Being conservative and extremely careful by nature, he was reluctant to make statements which were unsubstantiated. He felt confident, however, that eventually the whole problem would be solved and that the solution, when it came, would bring a new understanding of the mysteries of the laws of nature.

As Planck himself said:

> *What happens to the energy after emission? Does it spread out in all directions with propagation in the sense of Huygens' wave theory, in boundless*

progressive attenuation? Or does it fly out in one direction in the sense of Newton's emanation theory?

Be that as it may, in any case no doubt can arise that science will master the dilemma, serious as it is, and that which appears today so unsatisfactory will in fact eventually, seen from a higher vantage point, be distinguished by its special harmony and simplicity.

The Quantum Condition in the Everyday World

We do not notice the effect of Planck's quantum condition in the everyday world because the magnitude of quantum jumps is so small. For example, the energy of the pendulum of a pendulum clock is typically about 0.01 J and its frequency about 1 oscillation per second. This means that the energy of each quantum $hf = 6.63 \times 10^{-34}$ J and there are more than 1.5×10^{31} energy levels. The pendulum loses energy by cascading down these steps, which are so tiny that they are indistinguishable from a smooth, continuous slide.

Pendulum Clock.
Courtesy of Kilcroney
Furniture, Co.
Wicklow, Ireland

Chapter 5

Light — Wave or Projectile?

Max Planck applied his quantum theory to the energy of the oscillating charges which produce light and other electromagnetic radiation. He did not draw any conclusions about the light itself. In his Nobel speech he had posed

the question of whether the light would spread out in all directions, like a wave, or fly out like a projectile. Quanta of energy look more like projectiles than waves. This is hardly enough to dismiss the wave theory but still good reason to review the evidence that light is a wave. Does light have any properties which can be attributed to waves alone?

Waves Interfere with One Another

When waves meet, they interact in a characteristic way. Take two ships sailing close together; the water between them is churned up by the waves from the ships. In the photograph below, showing one ship being re-fuelled by another, we can see waves from the two ships moving towards each other at the front. They overlap towards the back of the picture causing a general disturbance. In this area when crests of two waves coincide they create larger crests, similarly troughs combine to make deeper troughs. When a crest meets a trough, they tend to cancel out; cancellation is complete if the two waves are of equal amplitude. These effects are

Photo # 19-N-294654   USS Kaskaskia refueling USS Hart, December 1944

USS Kaskaskia refuelling USS Hart. December 1944. Courtesy of US Navy

typical of waves and are called respectively, *constructive* and *destructive interference*. In this example the interference keeps changing from place to place, resulting in what looks like a random disturbance from which there is little to be learned about the nature of the waves.

Interference Can Create 'Stationary' Waves

It is possible to create a system of waves which continuously cancel one another in certain places, or along certain lines, by an arrangement involving waves with the same frequency, wavelength and amplitude. Such waves are called *stationary* waves.

The sources of the waves must be synchronized or *coherent*, which means that the waves do not change phase independently — we might think of them as dancers, who are dancing in step and can only change step if they all do so at the same time.

We can produce a really nice example of stationary waves on the surface of the water in a shallow glass tank by using two dipper sticks vibrating in phase. As the tips of the sticks enter the water they generate circular waves which spread out in all directions.

The image on the right is a photograph of such waves. The arrows show the directions of the so-called 'nodal' lines along which the waves cancel (interfere destructively). As time goes on, the two waves continue to spread out but the nodal lines remain fixed in space. The water remains undisturbed along these lines,

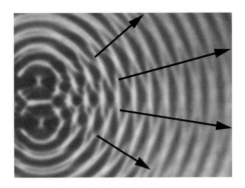

Water waves. Courtesy of Chris Phillips, Physics Department, Imperial College London

just as if there were no waves. In the areas between the lines there is *constructive interference* and the water oscillates up and down with double the amplitude while the overall pattern remains unchanged.

Stationary Waves of Light

Stationary waves in water are relatively common and not difficult to produce. It would be much more impressive to produce a stationary interference pattern with light, if light is indeed a wave. If this could be arranged, the two light beams would actually cancel along the nodal lines producing darkness!

Thomas Young (1773–1829) demonstrated just such interference of light in an historic lecture to the Royal Society of London in 1803. He inserted a thin piece of card edgeways on into a beam of sunlight, effectively splitting it into two and creating two coherent sources of light. The coloured bands seen on a screen placed behind the beam were evidence of constructive interference of the various wavelengths of light in the sun's spectrum. The dark lines between the bands were due to destructive interference. Those dark lines were the critical point of the demonstration; along them light waves were interfering destructively, effectively cancelling each other out at all times.

The demonstration was made complete by showing that when the light from one source was 'blocked off', the dark fringes vanished; where there had been darkness on the screen, there was now light! The

only logical explanation was that the act of blocking off the light from one side of the card was equivalent to removing one of the two sources in the water tank. This explanation could only be correct if light was behaving like a wave. To the audience of eminent scientists, Young said:

> *It will not be denied by the most prejudiced, that the fringes [which are observed] are produced by the interference of two portions of light.*

The following year, Young published a more detailed and quantitative account of the phenomenon in a paper entitled *Experiments and Calculations relative to Physical Optics.*[1] He subsequently produced interference between two point sources of monochromatic light, the optical equivalent of the water tank experiment. The figure

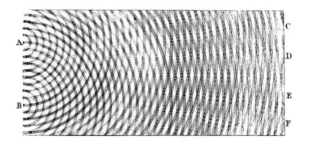

Young's sketch showing nodal lines

---

[1] Proceedings of the Royal Society of London A94, pp. 1–16. 1804.

above is a reproduction of an original sketch made by Young illustrating how waves from two point sources at A and B interfere destructively and give rise to dark bands C, D, E and F. The hypothesis that light is a wave soon became firmly established.

Thomas Young had a wide range of interests. Born in 1773 in Somerset, England, he was an infant prodigy, able to read at the age of two. He was an accomplished scholar of Greek, Latin, Arabic and Turkish and had an enormous interest in natural philosophy (physics) and natural history. He later entered university to study medicine, graduated and opened a practice in London. Having inherited a large sum of money from his uncle Sir Richard Brocklesby, he became a person of independent means and travelled to Egypt, where he deciphered the hieroglyphic script on the famous Rosetta Stone. At various times he was foreign secretary to the Royal Society of London, editor of the *Nautical Almanac,* and physician to the Royal Palladium Insurance Company.

Another Aspect of the Wave Nature of Light

In 1865, Maxwell had predicted that when an electric charge is accelerated it creates varying electric and magnetic fields, which continually regenerate one another and propagate through space at a definite speed. If the charge continues vibrating regularly, the result is an *electromagnetic wave* similar to the wave generated by a vibrating dipper stick in a water tank. Maxwell was able to calculate the speed of such a wave, and found that it was almost exactly the same as the speed of light, measured in 1849 by the French physicist Hippolyte Fizeau (1819–1896). This could hardly be a coincidence! In Maxwell's own words:

We can scarcely avoid the conclusion that light consists of the transverse undulations of the same medium which is the cause of electric and magnetic phenomena.

It was not until 1887 that Heinrich Hertz (1857–1894) demonstrated the existence of Maxwell's waves using a spark gap generator (an electrical apparatus which produces a spark across a narrow gap). The gap

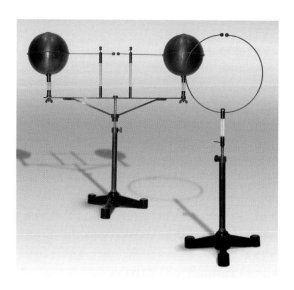

Spark gap generator and detector. Courtesy of John Jenkins, Spark Museum

is connected across an electrical circuit which bounces the spark to and fro, thousands of times per second, creating a very rapidly varying electric current and, according to Maxwell's theory, electromagnetic waves. To detect the waves, Hertz placed a wire ring connected to another spark gap a few metres away. A spark jumping across the first gap caused a second spark to jump across the gap in the ring. Hertz had generated and detected what we now call radio waves. He showed that the waves travelled in straight lines, that they are reflected and refracted in the same way as light and travel at the speed of light; in fact that light is an electromagnetic wave.

An Accidental Observation — The Photoelectric Effect

The secondary spark was difficult to see, so Hertz put the loop into a box to reduce stray light. To his surprise, he found that the spark,

far from being easier to see, became fainter and shorter. He was intrigued by the effect and began to suspect that some sort of light from the primary spark enhanced the brightness of the secondary spark. In a series of experiments, Hertz confirmed that light can even cause sparks and concluded (incorrectly) that the effect was due to ultraviolet rather than visible light.

Hertz published his results in a series of papers in *Annalen der Physik*, between 1887 and 1890, without commenting as to the possible reason for the 'effect':

> *I confine myself at present to communicating the results obtained, without attempting any theory respecting the manner in which the observed phenomena are brought about.*

Hertz did not realize that his discovery was of any practical importance. In reply to a question regarding possible applications, he is said to have replied:

> *No use whatsoever, this is just an experiment which proves Maxwell was right, we have mysterious electromagnetic waves which we cannot see with the naked eye, but they are there.*

Guglielmo Marconi (1874–1937), who had heard of Hertz's experiments as a teenager, recognized the potential of the 'mysterious electromagnetic waves'. Experimenting on his father's estate in Italy in 1895, he succeeded in sending wireless signals over a distance of more than 2 km. In 1899, he established radio

Gugliemo Marconi. Courtesy of An Post, Irish Post Office

communication across the English Channel and two years later, sent the first wireless signals from Europe to America.

Sadly, Hertz did not live to see that in his search for electromagnetic waves he had also stumbled across 'electromagnetic projectiles'.

Light and Electricity — The Investigation Continues

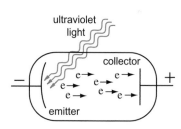

Hertz apparatus

Philipp von Lenard (1862–1947) was Hertz's assistant at the Bonn Institute from 1881–1884. About five years after the premature death of his one-time mentor, he began to investigate the photoelectric effect. Lenard mounted a clean metal surface (the *emitter*) at one end of an evacuated glass tube and a second metal plate (the *collector*) at the opposite end. These two electrodes were connected to a circuit. When the emitter was illuminated with ultraviolet light, a current was registered in the circuit. Lenard showed that the current passing through the vacuum was made of electrons ejected from the emitter and pulled across the tube by positive charge on the collector. He now knew why his former boss, Heinrich Hertz, had got fewer sparks when he put his apparatus in a box, inadvertently shielding it from illumination.

Lenard learned much more than that; he made the very important discovery that electrons were emitted with energies which depended on the *wavelength* of the light and not at all on its *intensity*. The emission of electrons seemed to be instantaneous.

Somehow, light seemed to liberate electrons from inside a metal. It was really surprising that it could produce such an effect. Does it mean, for example, that a copper coin exposed to sunlight emits electrons? The equations just did not add up; the power available was much too small. Assuming the light energy is spread out evenly and smoothly, as in a wave, the time required for an electron to receive the necessary energy would be about four months. But the effect was immediate, even if the light was very dim.

In the following years Lenard continued to work extensively in a variety of fundamental areas of physics. He demonstrated that the *cathode rays* liberated from a metal surface by heat or light are identical to electrons, which had been discovered by J.J. Thomson in 1897. His results were published in 1902 in *Annalen der Physik* and led to the award of the Nobel Prize in 1905.

In his Nobel speech, Lenard described both his work on cathode rays and his experiments on the photoelectric effect. Concerning the velocity of the photoelectrons, Lenard drew the following (incorrect) conclusion:

> *that the energy at escape does not come from the light at all, but from the interior of the particular atom. The light only has an initiating action, rather like that of the fuse in firing a loaded gun.*

The results of the photoelectric experiments were not correctly interpreted until 1905, when Einstein developed his theory of photons and produced a quantitative law which explained the results of photoelectric experiments. Lenard's paper was one of only four papers cited by Einstein in 1905 and referred to as '*the pioneering paper by Mr Lenard*'. However, 'Mr Lenard' felt he was disregarded and never forgave Einstein for attaching only his own name to the law.

The controversy can be considered in the context that Lenard was later to become a committed follower of the Nazi regime and a supporter of the idea that Germany should rely on 'Deutsche Physik' and ignore the misleading ideas of 'Jewish Physics'. He considered Einstein's theory of relativity a 'Jewish fraud'. With such credentials he was well fitted to become '*Chief of Aryan Physics*' and scientific advisor to Adolf Hitler.

Chapter 6

Einstein Enters the Scene

Born on 14 March 1879, Albert Einstein was 21 years younger than Planck. When he was still an infant, his parents moved from Ulm to Munich where he received his early schooling. Young Einstein was a bright child, interested in the world around him. When he was about 12 years old he derived, on his own, a proof of the theorem of Pythagoras and taught himself calculus at the age of 14. In school he disliked learning by rote and trying to memorize a series of dull facts. In fact he was anything but an ideal pupil, prone to sitting with a bored expression, gazing vacantly into the distance. Not surprisingly, this did not go down well with teachers who were accustomed to a rigorous German school discipline. One teacher has been quoted as saying: '*your very presence spoilt the respect of the class for me*'.

When his parents left Munich, Einstein obtained a medical certificate to the effect that he was suffering from nervous exhaustion. He left school at the end of fifth grade, and followed his family to Italy, where he received no formal schooling. As an alternative to having graduated from school, he sat the examination for entrance to the prestigious Swiss Federal Polytechnic (Eidgenössische Technische Hochschule, ETH) in Zürich. He failed at his first attempt; mathematics and physics presented no problem to the young Einstein, but his knowledge of languages and biology failed to satisfy the examiners.

Einstein spent a year at a school in Aurau in Switzerland, got his matriculation and entered ETH in 1896 but, true to character, even there he did not follow conventional protocol. Just as in school, the curriculum bored him; he missed lectures and paid little attention to the prescribed course. Fortunately a fellow student Marcel

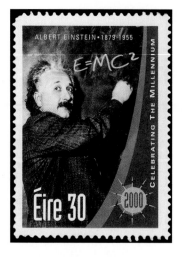

*Albert Einstein. Courtesy of
An Post, Irish Post Office*

Grossmann (1878–1936), who was to remain his good friend for life, took good notes and lent them to Einstein to cram before exams.

Following his graduation as teacher of mathematics and physics, Einstein spent two years in temporary jobs as a schoolteacher until, in 1902, he obtained a permanent position as an assistant in the patent office at Berne, where his duties included the examination of patent applications for electromagnetic devices. In his spare time he kept abreast with the latest developments in theoretical physics.

'Annus Mirabilis' 1905

In 1905, Einstein published no fewer than four papers in the prestigious journal *Annalen der Physik*. In them he addressed the fundamental unsolved problems of the time. When the Nobel Prize was awarded to him in 1921, the committee had an abundance of choices; their main difficulty being that, even then, the topics were controversial; many physicists found it difficult to accept Einstein's revolutionary new ideas. The papers (not in chronological order) were as follows.

On the existence of molecules

In August 1905,[1] Einstein published a paper entitled: *On the motion of small particles suspended in liquids at rest required by the molecular-kinetic theory of heat.* It was an extension of his doctoral dissertation, which he had presented to the University of Zürich. As he recalled

---

[1] Annalen der Physik 17, 549–560 (1905).

later, his intention was to look for '*facts that would guarantee as much as possible the existence of atoms of definite and finite size*'.

Brownian motion

In 1827, a Scottish botanist named Robert Brown (1773–1858) had noticed that tiny pollen grains suspended in water appeared to jiggle about randomly when viewed under a microscope. Einstein postulated that this is due to the thermal motion of water molecules, which bombard the pollen grain from all directions, as shown in the illustration. Millions of such collisions produce tiny displacements of the pollen, causing it to move in a pattern known as the '*random walk*', which may be analysed mathematically.

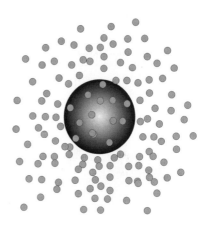

Brownian motion — pollen grain bombarded by water molecules

Einstein calculated that the mean displacement per minute of particles of diameter 1 μm² would be about 6 μm. His calculations confirmed that by looking through an ordinary microscope one may see evidence of the existence of water molecules which are about 10,000 times smaller than the grains of pollen.

A revolution in our notions of space and time — special relativity

In a paper published on 30 June 1905,[3] Einstein presented a completely new concept of space and time. By logical argument he postulated that there is no absolute frame of reference in the

---

[2] 1 μm or 1 micron = 10^{-6} m.

[3] *On the electrodynamics of moving bodies.* Annalen der physik 17, 891–921 (1905).

universe; the hypothesis of an 'ether', with respect to which one can define absolute motion, is then superfluous. Neither is there an absolute 'clock' to define the passage of time. Michelson and Morley had found no ether wind because there is no ether; they could now stop looking.

Redundant piece

The consequences of Einstein's logic are far-reaching. The piece representing the ether does not belong to the jigsaw and can be thrown away. Instead, there are a number of new pieces which fit together so beautifully that there can be no other explanation. Space and time are relative in the sense that they are components of a single complex entity. Time, which according to Isaac Newton: '*Of itself and by its own nature flows without relation to anything external*', is in fact related to space in an intrinsic way. The consequences of the relativity postulates are contrary to common experience; the most striking feature is that time is not the same for all observers. In principle, this could be verified experimentally, though at that time the necessary techniques were not available.

Mass-energy equivalence

On 27 September 1905, Einstein published yet another paper[4] in which he developed the theory of special relativity and deduced his famous equation $E = mc^2$. He had shown, not only that matter and

---

[4] *Does the inertia of a body depend on its energy content?* Annalen der Physik 18, 639–641 (1905).

energy are equivalent, but also that there is a precise relationship between them. As with his conclusions about space and time there was no means of experimentally testing the relationship, and again the statement was contrary to common perception. Einstein had to wait for experimental verification of his equation until 1932, when John Cockcroft (1897–1967) and Ernest Walton (1903–1995) working at the Cavendish Laboratory, split lithium nuclei into pairs of helium nuclei. As it happened, Einstein was in Cambridge at that time and was given a demonstration of the experiment.

Radical new ideas about light

Let us now go back six months to 17 March, when Einstein published his first paper on quantum theory. He addressed the problem of black-body radiation in a paper entitled: *On a heuristic[5] viewpoint concerning the production and transformation of light.* This paper is the most relevant to the Quantum Adventure, and the one for which he was awarded the Nobel Prize. Einstein's own opinion of this work is found in a letter written to his friend Conrad Habicht on 18 May 1905, in which he states: '...(it) *deals with radiation and the energy properties of light and is very revolutionary*'.[6] It is the only paper he describes in this way.

Einstein's Theory of the Photoelectric Effect

Einstein was aware of Planck's hypothesis of 'quantum bundles' of light; but this alone was not enough to explain the phenomenon of the photoelectric effect. In order to expel an electron from a metallic surface, the quantum of energy had to be localized in a tiny volume so that whatever electron was hit would receive the total quantum all at once. Einstein not only took up Planck's quantum

---

[5] The word '*heuristic*' implied that the matter was not necessarily in its final form and that alternative explanations were possible.
[6] *The Collected Papers of Albert Einstein*. Volume 5. Princeton University Press, New Jersey. 1995.

hypothesis, but went a step further to assume that the quanta of energy are localized in space, and behave just like particles, 'atoms of light', or *photons.*[7]

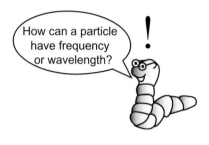

It is pertinent to ask how we can talk about the photon as a particle, and then, in the same breath, refer to its frequency. The new idea introduced by Einstein is so revolutionary, that normal vocabulary becomes inadequate. Everyday language can only be applied to objects with what we consider to be self-consistent properties — a particle is a particle and a wave is a wave.

Quantum mechanics requires such a revision of concepts, that it has occupied the minds of natural philosophers for the whole of the 20th century, and may indeed occupy them for many years to come. The photon *simultaneously* has properties of a particle and a wave. (For the remainder of the book we shall describe the photon in terms of wave or particle characteristics, whichever is appropriate.)

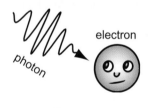

One on one

A characteristic property of metals is that the atoms have one or two outer electrons which are relatively free. Einstein reasoned that shining light on a metallic surface is equivalent to bombarding it with photons, some of which will collide with 'free' electrons near the surface. These collisions will be one-on-one 'all or nothing' interactions in which the photon will transfer its entire quantum of energy hf to the electron, all at once. If this energy is sufficient to overcome the forces which are pulling the electron back into the metal, it will join other 'liberated' electrons forming an *electron cloud* above the surface.

---

[7] Einstein did not use the word 'photon'. It is attributed to Gilbert N. Lewis, a Professor of Physical Chemistry at the University of Berkeley, California, who introduced it in a letter to *Nature* in December 1926.

Einstein's Equation

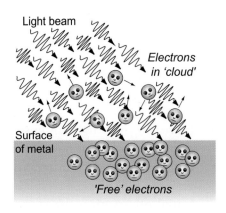

'Photoelectrons' are liberated

The liberated electrons will move about at random, some returning to the metal. Incoming photons in the light beam will liberate more electrons and equilibrium will soon be established between electrons entering and leaving the cloud.

Einstein deduced that the maximum energy of a liberated electron must be equal to the energy of the incoming photon (E_{max}) less the energy expended in overcoming the forces pulling it back into the metal. This escape energy, the '*work function*' *W*, refers to electrons nearest the surface and most easily liberated. Its value varies from one metal to another.

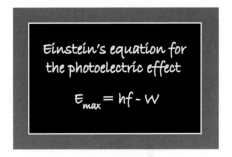

Einstein's equation for the photoelectric effect

$$E_{max} = hf - W$$

The photoelectric equation was mentioned specifically in the citation for the Nobel Prize. The equation itself is simple but the concept is revolutionary!

Consequences of Einstein's Equation

To establish experimentally whether or not Einstein's interpretation is correct we look for a number of essential features:

- The energy of the incoming photons (and therefore the frequency of the light) must exceed a certain threshold for the effect to occur. We might find, for example, that violet and ultraviolet photons expel electrons from a particular metal but red light has no effect at all.

- Above the threshold for emission, the energy of the energy of the most energetic photoelectrons should increase in proportion to the photon frequency.
- The number of electrons emitted per second should depend on the number of photons, i.e. on the intensity of the light, and nothing else.
- Increasing the light intensity should increase the number of photoelectrons but have no effect whatsoever on the energies.
- Emission should be instantaneous.

Initial observations by Hertz, Lenard and others were in general agreement with the above predictions but there was no quantitative experimental data to verify Einstein's theory with certainty. It was almost ten years before the results of experiments designed to test the validity of the equation were published. In the meantime, Einstein submitted his doctoral thesis and then occupied himself with other matters, such as Brownian motion and special relativity.

There is no evidence that Einstein doubted his light quantum hypothesis; he was probably quietly confident, in spite of strong resistance. Even Planck and Lorentz[8] expressed reservations to the suggestion that light quanta remain intact as they travel through a vacuum. Maxwell's theory of the propagation of light as a wave was difficult to relinquish.

Einstein is Proved Right

Meanwhile, across the Atlantic Ocean, in a laboratory at the University of Chicago, Robert Millikan (1868–1953) began a series

---

[8] Hendrik Lorentz (1853–1928) a Dutch physicist who shared the 1902 Nobel Prize for the discovery and theoretical explanation of the Zeeman effect.

of experiments designed to obtain more accurate data on the photoelectric effect. Millikan, already well known for the pioneering 'oil drop experiment' in which he had measured the electron charge, was a firm advocate of the wave theory of light. He was sceptical of Einstein's quantum theory and planned his experiments with the aim of disproving, rather than confirming, the experiment. Millikan's experiments were carried out very systematically and carefully. He was working with a definite goal in mind, and that goal was to test all aspects of Einstein's equation.

Millikan's apparatus was similar to that used by Lenard some years earlier. The emitter of a vacuum tube was made of a piece of clean, well-polished metal. When it was illuminated with light of certain wavelengths, electrons were emitted and attracted towards the positively charged collector.

Millikan now had a beam of photoelectrons 'in the open', in a highly evacuated tube, away from the forces of surrounding atoms and with no molecular collisions to worry about. As the accelerating voltage increased, the current increased likewise, until it reached its maximum, at which point the electrons were being swept across the tube as fast as they were emitted.

Measurements of this *saturation* current showed that it depends, as Einstein had predicted, solely on the intensity of the light and is independent of the potential difference between the emitter and collector.

An interesting effect was observed as the voltage was reduced. The current decreased accordingly, but surprisingly there was still a residual current in the circuit when there was no potential difference across the tube — nothing to pull the electrons towards the collector. The explanation, consistent with Einstein's theory, was that some photoelectrons are ejected with enough kinetic energy to carry them across to the collector without external help, although the majority eventually go 'back home' to the emitter.

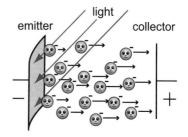

Electrons are pulled across the tube

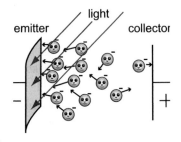

Electrons are driven back

Millikan then reversed the polarity of the electrodes, making the collector negative and emitter positive. What had been an accelerating potential was now a retarding potential. Even then, some electrons had enough kinetic energy to 'climb the potential hill'. He slowly increased the size of the retarding potential until even the most energetic electron was forced to return to base. From this, the *stopping potential,* Millikan was able to calculate the kinetic energy of the most energetic photoelectron. This electron would have come from the most favourable position near the surface of the metal — the condition assumed in Einstein's equation.

Millikan had announced his intention to measure the kinetic energy of electrons '*thrown out of a metal by ultraviolet light or X-rays*' at the Boston meeting of the American Physical Society in 1908, but while the principles of his method were straightforward, the practical details involved delicate measurements and the experiments took some years to complete. When he published his first results in 1914, he began his paper by elegantly describing the apparent contradictions introduced by Einstein's work:

> *in 1905 Einstein made the bold, not to say reckless hypothesis of an electromagnetic light corpuscle of energy hf, which energy was transferred upon absorption by an electron. This hypothesis may well be called reckless first because an electromagnetic disturbance which remains localized in space seems a violation of the very concept of an electromagnetic disturbance, and second because it flies in the face of the thoroughly established facts of interference.*

Millikan measured values of the maximum kinetic energy E_{max} at different frequencies of the incoming light, and repeated the measurements for different metals. He found that for any one metal, the data points lie on a straight line, showing that above the threshold frequency f_0, the maximum kinetic energy of the photoelectrons increases in direct proportion to the frequency. This confirmed the validity of Einstein's equation. The slope of the graph provided an independent estimate of Planck's constant, which was in good agreement with Planck's own value.

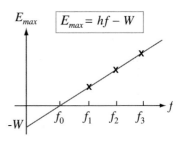

Einstein's equation is correct!

The 'work function' W varies from one element to another. Alkali metals such as caesium, potassium and sodium have low work functions and visible light quanta have enough energy to cause photoelectric emission, whereas metals such as copper need ultraviolet light. In our example of the copper coin, the ultraviolet component of sunlight will have a definite, if limited, effect.

Although apparently still not 100% convinced, Millikan sportingly conceded: *'This is a bullet-like, not a wave-like effect'*. His main

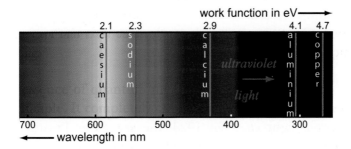

Work functions of some metals and the minimum wavelengths of light which liberate their electrons

paper, published in *Physical Review* in March 1916 has the unambiguous title: *A Direct Photoelectric Determination of Planck's 'h'.* Einstein's 'reckless hypothesis' had been proven correct, but the mystery of the wave-particle paradox had become greater than ever!

Millikan was awarded the Nobel Prize in 1923 for his work on the elementary charge of electricity and the photoelectric effect.

These words from his Nobel lecture reflect his continuing ambivalence about Einstein's work:

> *Einstein's equation is, I think, now universally conceded ... But the conception of localized light-quanta out of which Einstein got his equation must still be regarded as far from being established.*

Einstein was awarded the Nobel Prize in 1921 '*for his services to Theoretical Physics, and especially for his discovery of the law of the photoelectric effect.*' The citation is, at the very least, cautious and reflects the continuing uncertainty about much of Einstein's work.

The Compton Effect

In discussions of the photoelectric effect, the *momentum* of the photon has not come into play. This is because the target electron is 'attached' to the metal which absorbs all the momentum. The situation is analogous to a tennis ball which, when bounced off a wall, comes back with practically no loss of kinetic energy but with its momentum reversed.[9]

The Ultimate Evidence — Real Projectiles Have Momentum

An even more convincing experimental proof of the quantum behaviour of light came in 1923, the year after Einstein got his Nobel Prize. Arthur H. Compton (1892–1962) showed that the photon not only has energy and is localized in space but also, like a genuine particle, has momentum. In order to study collisions of photons with electrons which are essentially 'free', Compton used X-rays rather than visible light. An X-ray photon is over 1,000 times more energetic, making the force which binds the target electron to

---

[9] This analogy is by no means perfect — it applies to the backward momentum of the tennis ball, but in the case of the photoelectric effect, that momentum is transferred to the electron.

the metal insignificant and the process effectively a free particle collision. There is nothing to absorb the momentum. All participants are 'in the open' and both energy and momentum must be seen to balance.

In the diagram, an X-ray photon collides with a loosely held electron in a material such as graphite. It looks much like a classical collision between a moving and a stationary billiard ball, but there is an important difference. In the classical case, the incoming particle transfers some of its momentum to the target and therefore slows

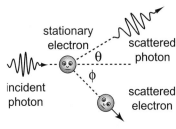

The Compton effect

down. Here the photon cannot slow down; photons always travel at the speed of light — a universal constant. Instead it changes its nature and becomes a photon of lower frequency and longer wavelength.

Whether in classical or in quantum mechanics, momentum has to balance, both in the forward direction and sideways. When combined with conservation of energy, this sets constraints on the dynamics of the collision. If the momentum and scattering

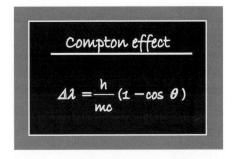

angle of the incoming photon are known, all the other parameters can be calculated.

Compton assumed that the momentum of a photon is $p = h/\lambda$ (the formula derived by Einstein 20 years earlier) and calculated the change in wavelength $\Delta\lambda$ of a photon which collides with a stationary electron in terms of the angle by which the photon changes direction (the scattering angle).

According to this equation, the wavelength change is the same for all incident radiation and depends only on the scattering angle. The percentage change is very small for visible light but becomes significant for X-rays.

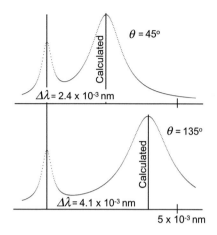

$\theta = 45°$

Calculated

$\Delta\lambda = 2.4 \times 10^{-3}$ nm

$\theta = 135°$

Calculated

$\Delta\lambda = 4.1 \times 10^{-3}$ nm

5×10^{-3} nm

Compton's results confirm Einstein's theory

Compton sent a beam of X-rays with a well-defined wavelength through a metal foil and measured the wavelengths of the outgoing radiation. The spectrum of the outgoing beam has two components, one with the same wavelength as the primary beam and the other with a longer wavelength. The value of that longer wavelength depends only on the scattering angle. The diagrams show the remarkable agreement between the experimental curves and the calculated values.

Arthur Holly Compton. Courtesy of NASA

As the photon is deflected through an angle θ, the 'free' electron recoils at an angle φ. In a subsequent experiment Compton looked for tracks of electrons at the predicted angle using an *expansion chamber*.[10] He found electron tracks at angles in close agreement with the formula, confirming that both energy and momentum is conserved in the process.

[10] Apparatus invented by C.T.R. Wilson (1869–1959) to make visible tracks of charged particles.

Compton received the Nobel Prize in 1927. In his acceptance lecture, he concluded:

We are thus confronted with the dilemma of having before us convincing evidence that radiation consists of waves, and at the same time that it consists of corpuscles.

Curiously, neither in the text of the lecture nor in the references, do we find mention of Albert Einstein.

Chapter 7

Niels Bohr Introduces the Quantum into Atomic Physics

Niels Bohr (1885–1962) was born in Copenhagen. He was awarded a fellowship by the Carlsberg foundation on completion of his PhD at the University of Copenhagen in 1911 and, like Rutherford before him, went to study with J.J. Thomson at the Cavendish Laboratory in Cambridge. On-going research there was focussed on the spectra of light from atoms, but explanations in terms of Thomson's 'plum pudding' model proved disappointing. The newly-arrived Bohr was outspoken and criticized aspects of Thomson's atomic model. This seems to have alienated the eminent man and '*J.J. politely indicated that it might be nice if he* [Bohr] *left Cambridge and went to work with Rutherford*'.[1]

Bohr Moves to Manchester

At Rutherford's laboratory in Manchester, research centred on alpha particles and their effects on matter. It was less than a year since Rutherford had established his model of atomic structure and Manchester was an exciting place to be. Rutherford was very much 'hands-on'; an inspired and enthusiastic leader who was genuinely interested in his students and was always ready to discuss any detail of an experiment, no matter how small. The young Bohr was a theorist at

---

[1] E.U. Condon. *60 years of quantum mechanics*. Physics Today 15, 45 (1962).

heart, although he enjoyed using his hands and had even won a gold medal in Denmark for some of his experiments. Asked why he was so readily (and uncharacteristically) impressed by a theorist, Rutherford replied:

'*Bohr's different. He's a football player!*' Niels and his brother Harald had played football for one of the top amateur clubs in Denmark. Harald was selected to play for the Danish national team which won a silver medal in the 1908 Olympic Games in London.

By that time, it had been established beyond all doubt that the atom was mostly empty space, with the positive charge and almost all the mass concentrated in a tiny nucleus. The negatively charged electrons had to be arranged in some way about the nucleus but the nature of the 'arrangement' remained an insurmountable problem.

If electrons orbit the nucleus in a planetary fashion, they should, according to classical physics, emit light continuously. As a result they would lose energy and spiral into the nucleus. If, on the other hand, they are stationary, they should fall into the nucleus like apples from a tree. In either case, the atom would be unstable. Rutherford was well aware of this, but chose to put it aside. He left further exploration of atomic structure to others.

Bohr Takes up the Challenge

To explain how the atom remains stable formed an ideal challenge for 26-year-old Niels Bohr. He set out to formulate a model of the simplest atom, hydrogen, with just one electron. Having exhausted all other options, he came to the conclusion that when an atom is in its natural rest state, the electron must be in a special orbit, a 'stationary state', to which the normal rules do not apply. He had to find some reason why an electron in one of these stationary states should not emit light.

A New Element and a New Idea

While looking at the experimental information on atomic spectra from all sources, Bohr's attention was caught by a series of papers by the Cambridge astrophysicist, J.W. Nicholson (1881–1955) concerning a theoretical model to explain spectral lines observed almost 50 years earlier in light from a planetary nebula.[2] These were first reported by William Huggins (1824–1910), an amateur astronomer who worked from his home in the suburbs of London. Huggins was the first person to examine the absorption spectra from stars other than our sun, and wrote:

On the evening of the 29th of August, 1864, I directed the telescope for the first time to a planetary nebula in Draco (NGC 6543). The reader may now

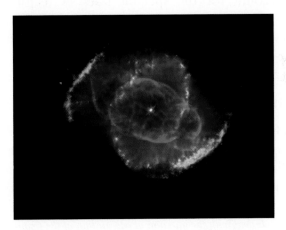

NGC 6543 Cat's eye nebula. Courtesy of NASA/AURA/STSCI

be able to picture to himself to some extent the feeling of excited suspense, mingled with a degree of awe, with which, after a few moments of hesita-tion, I put my eye to the spectroscope. Was I not about to look into a secret place of creation? I looked into the spectroscope. No spectrum such as I expected! A single bright line only![3]

---

[2] In the 18th century, astronomical objects other than planets and comets were called nebulae (from the Latin word for cloud). Planetary nebulae are a subgroup, so named by astronomer William Herschel (1738–1822) because they resembled large planets such as Uranus, which Herschel had discovered in 1871.

[3] William Huggins. *The New Astronomy: A Personal Retrospect.* The Nineteenth Century, 41, 1897. pp. 907–29.

Huggins realized that the bright green line was part of the emission spectrum of a glowing gas. It did not belong to the spectrum of any known element so he invented a hypothetical element called 'nebulium'. Unidentified lines in the emission spectra from similar nebulae were soon discovered and presumed to be associated with the new element.

There was no sign of nebulium anywhere on Earth and, in 1911, Nicholson came up with another explanation. He devised a complicated model based on 'exotic oscillations' of electrons of known elements to account for the spectral lines. A feature of special significance was that he introduced Planck's quantum of action to get correct values for the wavelengths. It turned out that his model was wrong — but the idea was right!

Bohr Takes up the Challenge

After just a few months at Manchester, the term of Bohr's fellowship came to a close and he went back to Copenhagen to become a physics lecturer. He had become aware of Nicholson's work in late 1912 but, as he wrote in a letter to Rutherford at the end of January 1913:

> *I do not at all deal with the question of calculation of the frequencies corresponding to the lines in the visible spectrum ...*

In Bohr's view, he and Nicholson were looking at different things. Spectra gave information about how energy was emitted as the atom settled into states of lower energy. Bohr was interested only in the natural rest state (*ground state*) of the atom.

Bohr Changes His Mind

Less than a month after Bohr had written to Rutherford dismissing the relevance of spectra, a colleague drew Bohr's attention to Balmer's formula for visible hydrogen lines. The formula had been re-written by Johannes Rydberg (1854–1919) to give the frequencies, rather than the wavelengths, of lines. The new format looks quite different and Balmer's constant has been replaced by other constants (R and c) but the positions of the lines are given by putting $n = 3, 4, 5,...$ in exactly the same way as before.

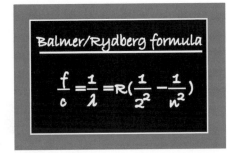

Bohr's opinion of spectra changed almost overnight — in his own words: '*As soon as I saw Balmer's formula, the whole thing was immediately clear to me*'.

The Significance of Balmer's Formula

According to Balmer's formula, the frequencies depend on the difference between two numbers. The numbers themselves are not unique to a particular line but recur on a regular basis and the pattern is best illustrated by arranging them in a table.

series	frequency values						
Lyman series	A-B	A-C	A-D	A-E	etc		
Balmer series		B-C	B-D	B-E	B-F	etc	
Paschen series			C-D	C-E	C-F	C-G	etc

Interpretation of the frequency values

The numbers are very large and irregular so we will denote them by the letters (A, B, C, D, E etc.). The spectral line frequencies fall into distinct groups or *series*, corresponding to the value of the first letter.[4]

---

[4] Each series is named after the scientist who discovered it.

In the Balmer series, the frequency of the red line corresponds to B-C, the blue line to B-D and the purple lines to B-E and B-F. The spacing between the lines decreases rapidly and the frequencies converge

Balmer series of spectral lines

to a value which is just inside the ultraviolet part of the spectrum. The other series are in other non-visible parts of the spectrum.

When Niels Bohr saw Balmer's formula, he realized that the numbers we have labelled A, B, C etc. are related to values of atomic energy and that electromagnetic radiation is emitted as a result of abrupt changes in that energy. Planck had made the hypothesis that an oscillator gains or loses energy in *quanta* of size $E = hf$, corresponding to the difference between two allowed energies. Perhaps a similar restriction applies also to atoms?

Bohr Invokes Quantum Theory

In July 1913, only a matter of months after he had seen Balmer's formula, Bohr published details of a revolutionary model of the hydrogen atom. He started by referring to the alpha particle scattering experiments and to the models of Rutherford and Thomson. He concluded by saying:

> *Whatever the alteration in the laws of motion of the electrons may be, it seems necessary to introduce in the laws in question a quantity foreign to the classical electrodynamics, i.e. Planck's constant, or as it often is called the elementary quantum of action.*

Bohr developed the point in his Nobel speech of 1922:

> *It has, however, been possible to avoid the difficulties of the electrodynamical [Rutherford] model by introducing concepts borrowed from the so-called quantum theory, which marks a complete departure from the ideas that have hitherto been used for the explanation of natural phenomena.*

The Key to our Existence

Bohr's challenge had been to find the answer to the fundamental question: 'Why are atoms stable?' He found the answer in terms of Max Planck's *quantum of action* which imposes a new 'code of behaviour' on the atom. There are certain *stationary states* in which the atom can exist, which means that atomic energy is *quantized*. The emission and absorption of electromagnetic radiation occurs only as a result of transitions between stationary states. The frequency of the radiation is determined by the energy difference between the states, according to the quantum condition: $hf = E_2 - E_1$. Once the atom reaches the lowest quantum state, there is nowhere to go and it will remain in that state indefinitely, unless it acquires energy from outside. Planck's quantum ensures the stability of universe!

Energy Levels — A Simple Picture of the Atom

The most realistic way to describe atoms is in terms of what we can measure; the energies of stationary states and the frequencies of spectral lines.

Atoms spend most of their time in the ground state, where they have the least amount of energy. They can be promoted into a higher energy *excited* state by absorbing an amount of energy exactly equal to the energy difference between the two states.

Excited states are unstable and the atom very quickly returns to the ground state, emitting light at a frequency determined by Bohr's quantum condition.

According to Bohr's model, the energies of the stationary states — *energy levels* — have values which depend on the value of the *quantum number n*. The energy of the ground state corresponds to $n = 1$.

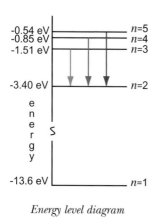

Energy level diagram

The energy levels, and also the allowed transitions which correspond to spectral lines, may be shown in an energy level diagram. The spacing between energy levels is larger at lower values of n (where the electron is closer to the nucleus).

As n increases, the energy levels become very close together, converging to a value which we choose to be $E = 0$, where the electron becomes free. Bohr's theory gives the energy needed to free the electron from a hydrogen atom (the *ionization energy*) as 13.6 eV, which corresponds to the measured value.

The diagram shows transitions which go to the $n = 2$ level. The left hand arrow corresponds to a transition from the adjacent $n = 3$

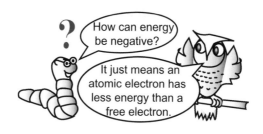

level; the remaining arrows are the result of transitions from higher energy levels. These transitions correspond to spectral lines in the Balmer series.

Visualizing the Bohr Atom

The energy level diagram gives an accurate description of the atom in terms of scientifically meaningful parameters. However, if we want to form an image it is not the first thing which comes to mind. The planetary model, in which electrons orbit the nucleus, is visually more satisfactory provided we treat it as an artistic impression rather than a working model. The fact that only certain energies are *allowed* implies that only certain values of frequency and radius are allowed. The diagram (right) shows the first few allowed orbits. When an atom is in the ground state, the electron is in the orbit closest to the nucleus (*the first Bohr orbit*); the size of that orbit defines the size of the atom. Here also, Bohr's theory was found to be in good agreement with experimental data.

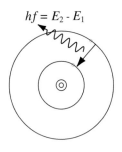

$$hf = E_2 - E_1$$

Artist's impression of electron orbits

Transitions between stationary states correspond to electrons making *quantum jumps* from one allowed orbit to another. An electron can jump from any excited level to the next highest level, or jump directly down several levels, hence the various series of spectral lines.

Seamless Transition From the Atomic to the Household World

If quantum mechanics is a proper representation of the laws of nature, it must hold everywhere; not only in the world of the atom, where quantum effects are dominant, but also in the household world, where they are negligible. Max Planck realized from early on that: '*The classical theory can simply be characterized by the fact that the quantum of action [h] becomes infinitesimally small*'.

Bohr used this principle as the 'acid test' of any formula in quantum mechanics; it must translate into the corresponding formula in classical physics when Planck's constant can be neglected. He called this the *correspondence principle*.

The principle is beautifully illustrated in the expression for the frequency of the radiation emitted by the Bohr atom, although it is not immediately obvious. The formula $hf = E_2 - E_1$

appears to bear no relation to the classical expression for the frequency of light emitted by an accelerated electric charge, so how can there be a smooth transition from quantum to classical physics?

Bohr described the orbital motion of atomic electrons using equations similar to those used by Newton to describe the orbit of the moon. In this classical picture, the electron emits light at a frequency corresponding to the number of orbits completed per second and depends, not on the difference between allowed energies, but on the energy of a specific orbit.

Bohr then showed that, for transitions between neighbouring allowed orbits at large values of n, the classical result was the same as that given by Planck's formula. Bohr needed the values of the mass and charge of the electron for his calculations and he paid special tribute to *'the beautiful investigations of Millikan'* through which the quantities e, m, and h are known.

In his Nobel lecture in 1922, Bohr describes the calculation in typical style. Every word chosen with great care, to convey precisely what he wanted to say; a language his colleagues and students often referred to as *'Bohrspeak'*:

> *It was possible to show that the frequency of radiation sent out during the transition between two stationary states, the difference of the term numbers of which is small in comparison with the term numbers themselves tended to coincide in frequency with one of the harmonic components into which the electron motion could be resolved and accordingly with the frequency of one of the wave terms in the radiation which would be emitted according to the laws of ordinary electrodynamics.*

Stationary States Exist!

Bohr's model was developed to describe the hydrogen atom and was most successful in that context. It gave correct values for spectral line frequencies, the size of the hydrogen atom and the ionization energy. But there was no experimental proof that stationary states really did exist.

Within a year, James Franck (1882–1964) and Gustav Hertz (1887–1975) unwittingly demonstrated that mercury atoms accept energy only in amounts corresponding to the energy difference between two stationary states. They were studying how electrons lose energy when they pass through a gas and were completely unaware of Bohr's model and its implications:

> *when we made the experiments we did not read the literature well enough ...*
> *On the other hand, one would think that other people might have told us*
> *about it ... we had a colloquium at the time in Berlin ... all the important*
> *papers were discussed. Nobody discussed Bohr's paper.*[5]

Franck and Hertz measured the current passing through a discharge tube filled with mercury vapour, as the voltage was increased. They recorded an abrupt drop in the current at 4.9 V. At that point, for some mysterious reason, the energy of almost all the electrons was being transferred to the mercury atoms. This behaviour was replicated at 9.8 V and 14.7 V (integer multiples of 4.9 V).

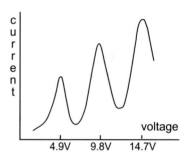

Evidence for stationary states

From the point of view of Bohr's theory, such behaviour is not mysterious at all. The mercury atoms are absorbing energy from electrons accelerated through 4.9 V to induce a transition between two energy levels separated by exactly 4.9 eV.

Bohr's theory was received favourably and got a 'good press', even to the point of being mentioned in *The London Times* newspaper. Albert Einstein described it as '*very remarkable*'. However, Bohr himself admitted that it was fundamentally incomplete; it offered no explanation as to *why* there are stationary states or *why* some transitions happen more frequently than others, as is indicated by variations in the intensity of spectral lines.

---

[5] Extract from an interview given by Franck in 1960.

Bohr 'broke the mould'. He introduced the quantum into atomic theory in a way which gave enough agreement with experiment observations to act as a wake-up call. He was searching, not for the answer to a specific question, but for the right questions to ask. His work was the catalyst for a whole new theory based on *quantum mechanics* but it took more than ten years, and the collective efforts of many brilliant minds, to create that new theory and Bohr played no small part in bringing these minds together and coordinating their efforts.

The Niels Bohr Institute

In 1916, Bohr became Professor of Physics in Copenhagen and began to lobby for the establishment of an institute for theoretical physics. University funding was not forthcoming but Bohr managed

The Bohr Institute in Copenhagen

to raise enough funds privately to purchase the site, despite the difficult economic climate. The construction work was state-funded and the institute was opened in 1921. The Carlsberg Foundation gave an annual grant to help fund research. Bohr was nominated director of the institute and had an apartment on the first floor.

He valued the contributions of young physicists, with their enthusiasm, fresh insights and sometimes critical attitudes.

The institute, with its extremely informal atmosphere attracted the cream of bright young physicists. They came from all corners of the world: Werner Heisenberg and Albert Einstein from Germany, Lise Meitner (1878–1968), Wolfgang Pauli and Erwin Schrödinger from Austria, Georg Hevesy (1885–1966) from Hungary, Hendrik Casimir (1909–2000) from the Netherlands, George Gamow and Lev Landau (1908–1968) from Russia, Robert Oppenheimer from the USA, Kazuhiko Nishijima (1926–2009) from Japan, Paul Dirac from England... The list is endless and contains the names of many future Nobel Prize winners and even some who were to become household names in physics.

Niels Bohr became a national figure in Denmark. He was the complete antithesis of the aloof academic. He was knowledgeable in art, history, the stock market and just about any human activity. Not only was Bohr a talented footballer, he was also a good skier and sailor. The physicists in his institute worked hard and played hard. They commandeered the library tables to play table tennis tournaments and sat at the same tables late into the night working on the emerging quantum mechanics. They had beer parties at the local taverns and wrote classic papers which would revolutionize physics.

On one occasion, when returning from a party, Bohr and the young Casimir and Gamow made a bet as to who could climb the outside wall of a bank in central Copenhagen. The police were called, but it seems were not unduly surprised; it was only Professor Bohr after all!

In 1929, the first of a unique series of conferences was hosted at the institute. The conferences were informal, the atmosphere was relaxed, and sometimes participants even provided entertainment, with a flavour of the topics currently being discussed.

Chapter 8

Werner Heisenberg — An Equation for Uncertainty

Heisenberg

Werner Heisenberg (1901–1976) was born in Würzburg, Germany on 5 December 1901, almost exactly one year after Planck had presented his quantum hypothesis to the German Physical Society. At the age of 18, he entered the University of Munich to study physics and mathematics under Arnold Sommerfeld (1868–1951). Soon it became clear that young Heisenberg needed to be stretched beyond the standard undergraduate curriculum, and Sommerfeld allowed him to attend research seminars for advanced students. At the first of these, Heisenberg *'spotted a dark-haired student with a somewhat secretive face in the third row ... His name was Wolfgang Pauli, and for the rest of his life he was to be a good friend, though often a very severe critic'.*[1]

Werner Heisenberg. Courtesy of the Nobel Foundation

Sommerfeld had a close association with Max Born (1882–1970), who was the Professor of Theoretical Physics at the University of

---

[1] Werner Heisenberg. *Physics and Beyond.* Allen & Unwin, London. 1971.

Göttingen. In June 1922, Sommerfeld brought Heisenberg to Göttingen, where Niels Bohr was giving a series of lectures on quantum physics. This was Heisenberg's first meeting with the person who, more than any other, was to influence his life. The meeting ignited in Heisenberg an instant fascination with the unexplored mysteries of the atomic world.

Bohr was quick to spot the most talented students and gave both Heisenberg and Pauli an open invitation to come to Copenhagen for a year. Heisenberg was thrilled: '*Suddenly the future looked full of hope and new possibilities*'.[1] But, before he could go, Heisenberg had to finish his doctorate thesis (on *'Turbulence'* — a completely unrelated topic) so his friend Wolfgang Pauli (1900–1958) went ahead without him.

There was another bonus. Heisenberg got the offer of a job as assistant to Max Born. On his return from Copenhagen, his employment at the Institute of Physics in Göttingen was assured.

Heisenberg and Pauli have Serious Misgivings

The challenging questions of the time, concerning the world of atoms, are echoed in a conversation between Heisenberg and Pauli, which took place around 1922:

> Pauli: *Do you honestly believe that such things as electron orbits really exist?*
> Heisenberg: *… we can observe the path of an electron in a cloud chamber … there is such a thing as an electron trajectory … and we can take it that it will appear in the atom as well. But I have some reservations on that score … And when it comes to electrons jumping from one orbit to the next — as the theory demands — we make sure not to specify whether they make high jumps, long jumps or some other sorts of jump. It all makes me think that something is radically wrong with the whole idea of electron orbits.*[1]

In March 1924, Heisenberg eventually got to Copenhagen; by then Pauli had been there for about 18 months. At first, he was rather intimidated by the other students, but quickly settled down as he got to know Bohr. The intellectual environment was ideal; he played a full part in the philosophical discussions and also contributed his own ideas.

The doubts that Heisenberg had expressed earlier to Pauli persisted; he felt that Bohr's model of the atom, while it presents an intuitive picture, could not be considered as a complete theory. It was what its name implied, just a model. We cannot 'see' an electron; it is so delicate that even the impact of a single photon changes its position and momentum. The act of 'seeing' would destroy what we are trying to observe.

Before 1925, quantum mechanics was a mixed bag of semi-classical recipes and hypotheses. Each problem was solved individually, because there was no coherent theory, and the emphasis was always on finding a solution which would reproduce the observations. The technique was first to find the classical solution and then somehow 'translate' it into the language of quanta. The quantum solution had to satisfy Bohr's correspondence principle and merge seamlessly with classical theory when moving from the world of the atom into the everyday world. Finding appropriate quantum solutions required skill, intuition and even guesswork; correspondence became something of an art.

In Copenhagen, Heisenberg became very familiar with the art of correspondence. When he returned to Göttingen, he tried to guess at formulae which would predict the intensities of lines in the hydrogen spectrum, but he found himself '*in an impenetrable mass of complicated mathematical equations with no way out*'.[1]

He was now sure that this whole process of concocting solutions, simply to reproduce experimental measurements, was flawed. Maybe he should turn the process on its head and start with what was already known?

By the spring of 1925, Heisenberg was convinced that he should not think in terms of electron orbits at all. As he explained at the Nobel ceremony in 1933:

Modern physics has definitely decided in favour of Plato. In fact the smallest units of matter are not physical objects in the ordinary sense; they are forms, ideas which can be expressed unambiguously only in mathematical language.

Signals from the Atoms

Light is nature's way of sending first hand information from the world of the atom. The frequencies and intensities of the spectral lines can be observed and measured. Now Heisenberg was about to use this information to construct the equations of a new mathematical theory.

Heisenberg's frequency array

Heisenberg arranged the frequencies methodically in the rows and columns of an array. The diagram shows the 'corner' of such an array.

The frequencies of the Balmer series, with its well known red, blue and purple lines are arranged in the appropriately coloured boxes. The emission frequencies appear in column two, and the corresponding absorption frequencies in row two. A similar scheme applies to frequencies of other series, which are not visible and therefore not so easily displayed! The grey diagonal squares do not correspond to transitions and contain zeros.

Working Without a Model

A model, even if it is not completely correct, makes it possible to visualize what one is doing. Now that Heisenberg had rejected the whole idea of electron orbits, he was left with arrays of numbers, devoid of an image to which they could be related. He was working 'in the dark' purely with the logic of mathematics; like a grandmaster playing chess blindfold, with no board or pieces to distract him.

As Lagrange had said in *Mécanique Analytique* almost 140 years earlier, there would be no diagrams. The same was true of

Heisenberg's work; but it also meant that his arguments were that much more difficult to follow.

Starting with a Simpler System

After he '*had come to grief*'[1] in his first attempt to construct a theory of the hydrogen atom based solely on the observed frequencies and intensities, Heisenberg decided to simplify matters by thinking of the atom as some sort of oscillator.

According to classical theory, an oscillating charge, such as an electron in an antenna, emits electromagnetic radiation at the frequency at which it vibrates. Heisenberg assumed that the same is true in the quantum world.

In Planck's theory, individual electrons can vibrate at any one of the frequencies corresponding to an allowed energy. The light forms a continuous band as each electron makes its contribution.

According to Heisenberg, vibrations at a set of frequencies characteristic of a particular atom and with different amplitudes, combine to form the spectrum of the atom.

A classical analogy is the human voice. The vocal chords have a characteristic minimum or *fundamental* frequency but can also vibrate at integer multiples or *harmonics* of that frequency. The pattern of relative strengths of the harmonics, the *harmonic profile*, is different for each individual and gives the voice its characteristic quality. The ear is very sensitive to harmonic profile; we can recognize the 'hello' of a friend on the telephone by the quality of their voice. In crime detection, electronic analysis of a recorded voice is used to break down individual sounds into their harmonic profiles, as a means of identification; this is known as 'voice printing'. Voice prints rely on *Fourier analysis*, a technique developed 300 years ago by Jean Baptiste Fourier (1768–1830).

The Voice Prints of an Atom

The process of Fourier analysis can be used in reverse to reconstruct a sound from the intensities of its components. An interesting challenge would be to reconstruct a model of the vocal chords and vocal cavity and perhaps even of the whole person from their voice print.

Classically, any time-dependent quantity can be expressed in terms of a Fourier series. Heisenberg decided to apply similar methods to quantum systems such as atoms. Thinking of the atom as a set of oscillators he could apply Fourier analysis to the motions of electrons, using the measured frequencies and intensities, without any reference to orbits. The end result would be a mathematical theory of the atom, constructed from its 'voice print'.

As he said in a lecture many years later:

> *The idea suggested itself that one should write down the mechanical laws not as equations for the positions and velocities of the electrons but as equations for the frequencies and amplitudes of their Fourier expansion.*[2]

We should not despair if all this sounds too confusing and complicated. Practically every one of Heisenberg's peers would find it so, even Born and Bohr. There was also the criticism that oscillators were just as unobservable as Bohr's electron orbits. Heisenberg himself faced unanswered questions and was by no means confident that he was right.

Heisenberg goes to Heligoland

Towards the end of May 1925, Heisenberg became ill with a serious attack of hay fever and had to ask Born for 14 days' leave of absence.

---

[2] Werner Heisenberg. *Physics and Philosophy: The Revolution in Modern Science.* Harper & Row, New York. 1958.

He went to the rocky, pollen-free island of Heligoland, in the North Sea, where he would be able to walk and swim at leisure. *'I must have looked quite a sight with my swollen face; in any case my landlady concluded I had been in a fight and promised to nurse me through the after effects'.*[1] Relaxed, refreshed and without distractions Heisenberg finally put together the pieces of his new mechanics.

The places in the story

All the numbers in an array, taken together, represent the value of some particular quantity, such as position, and have to be treated as a unit. To arrive at equations of motion, Heisenberg had to develop ad hoc methods of dealing with the arrays. Adding arrays was fairly straightforward — each number in one array was added to the corresponding number in the other array, and similarly for subtraction. He also devised a method of multiplying one array by another array, but it had a strange consequence; the order in which arrays were multiplied affected the result. The product of two arrays $A \times B$ was not the same as the reverse product $B \times A$.

Heisenberg Watches the Sunrise

The most outstanding unanswered question was: would energy be conserved in the new theory? The energy lost by the atom had to be the same as the energy carried away by the radiation. One evening, Heisenberg had reached the point of calculating the terms in the energy table. He was so excited when the first terms came out correctly that he began to make one arithmetical error after another, in

Heligoland

what was, by then, a complex series of calculations. At 3 am, he finally obtained his results and they were correct.

At first, I was deeply alarmed. I had the feeling that, through the surface of atomic phenomena, I was looking at a strangely beautiful interior … and so, as a new day dawned, I made for the southern tip of the island … to climb a rock jutting out into the sea … and waited for the sun to rise.[1]

The First Paper

Concerned about the meaning of the strange multiplication rule, Heisenberg went to Hamburg to see Pauli, before going back once more to Göttingen. Pauli, who could be relied on for a critical assessment, was enthusiastic. Heisenberg then wrote what was to be a ground-breaking paper and, on 9 July 1925, gave the manuscript entitled *Quantum mechanical reinterpretation of kinematic and mechanical relations*[3] to Born asking him either to 'throw it into the fire' or send it to *Zeitschrift für Physik* for publication. He (Heisenberg) had been

---

[3] Werner Heisenberg. *Über quantentheoretische Umdeutung kinematischer und mechanischer Beziehungen*. Zeitschrift für Physik 33, 879–893 (1925).

invited to lecture at the Cavendish Laboratory in Cambridge and departed for England, leaving his manuscript to its fate!

Born recognized the significance of Heisenberg's idea and almost immediately sent the manuscript to *Zeitschrift für Physik*. He wrote to Einstein: '*Heisenberg's latest paper, soon to be published, appears rather mystifying but is certainly true and profound*'.[4]

Born recalled at his Nobel lecture in 1954:

> *I could not take my mind off Heisenberg's multiplication rule, and after a week of intensive thought and trial I suddenly remembered an algebraic theory which I had learned from my teacher, Professor Jakob Rosanes, in Breslau.*

The mathematical method Jakob Rosanes (1842–1922) had taught was *matrix* theory, which mathematicians had devised, not for any particular purpose but, as an exercise in logic and for satisfaction. There were rules, similar to Heisenberg's, for multiplying one matrix by another, all consistent and logical, as is always the case in mathematics.

Rosanes was also a well-known chess player. In 1863, he played a seven-day match in Breslau against the German master Adolf Anderssen who was, at that time, considered to be the world's best player. The match ended in a draw.

The Role of Matrices in Quantum Theory

Heisenberg's arrays were identical to matrices. There would be a matrix for everything; a matrix for position, a matrix for momentum, a matrix for frequency, a matrix for the hydrogen atom, a matrix for an electron. Operating on the 'hydrogen matrix' with the 'frequency matrix' would give the observed lines of the hydrogen spectrum. Multiplying the matrix of an electron by the momentum matrix would give the momentum of the electron.

---

[4] Letter from Born to Einstein 15 July 1925.

There was also a definition and method for determining the *eigenvalues* or *proper values* of a matrix, a set of numbers which subsequently turned out to have an important physical significance.

A chance encounter brought another player into the Göttingen team.[5] Born was travelling to Hanover by train in the company of a colleague. In the course of a conversation about his work on the new quantum theory, Born said it was progressing quickly but that there were difficulties with the manipulation of matrices. A young student, Pascual Jordan (1902–1980), who happened to be in the same railway carriage, introduced himself to Born; he had experience in manipulating matrices and offered to help Born develop his theory.

Born readily accepted the offer and the theory of matrix mechanics developed quickly over the following months. While Heisenberg was still away, Born and Jordan wrote a paper entitled *On Quantum Mechanics*, which reached *Zeitschrift für Physik* on 27 September 1925, just 60 days after Heisenberg's first paper. The paper had developed Heisenberg's equations in matrix form.

Born and Jordan sent a pre-print of the paper to Heisenberg in Copenhagen. He confided in Bohr: '*I got a paper from Born, which I cannot understand at all. It is full of matrices, and I hardly know what they are*'.[6]

To Commute or Not to Commute

The strange rule Heisenberg had invented to multiply his arrays is also a feature of matrix algebra.

In normal algebra, *b* multiplied by *c* gives the same result as *c* multiplied by *b*. As every school child learns, $2 \times 3 = 3 \times 2$; the order does not matter. This is called the law of *commutative* multiplication. However, the *matrix* product **BA** does not generally give the same result as **AB**; matrix multiplication is *not* commutative.

---

[5] *Archive for the History of Quantum Mechanics.* Interview with Born. June 1960.
[6] Nancy Greenspan. *The End of the Certain World. The Life and Science of Max Born: The Nobel Physicist Who Ignited the Quantum Revolution.* John Wiley, Chichester. 2005.

In this new mechanics, the measurement of one property (*observable*) **A**, followed by the measurement of a second observable **B** is represented by the product **BA** ('**B** *after* **A**') and is not necessarily the same as **AB** ('**A** *after* **B**'). So, matrix mechanics was implying that the order in which physical parameters are measured may affect the result. Was this a major fault? Did it make the whole theory invalid?

Paul Dirac (1902–1984), who had also seen a pre-print of Heisenberg's original paper, quickly realized that far from presenting a problem the matrix equations were telling us something important. Another secret of nature was being revealed. The order in which the value of certain observables is determined *does* matter and is, in fact, a dominant feature of the world of atoms and subatomic particles.

Born Lays the Foundations for the Uncertainty Relation

It was then that Born made the critical contribution. Not only do the matrices **q**, representing *position*, and **p**, representing *momentum*, not

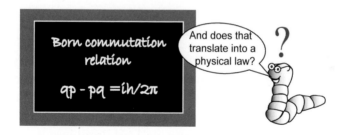

commute, but the result of the *commutation relation* is a number proportional to Planck's constant. To fit into the mathematics, Born had to add one further detail, an *imaginary* number *i* (the square root of minus one).

Born said later: '*I shall never forget the thrill I experienced when I succeeded in condensing Heisenberg's ideas on quantum conditions in the mysterious equation …*'[7]

---

[7] *Archive for the History of Quantum Physics*. Interview with Born. 1963.

The commutation relation implies that if we measure the *position* of an object after measuring its momentum, we will obtain a different result (within limits defined by Planck's constant) than if we measure the *momentum* after measuring the position.

Heisenberg, who had rapidly 'come up to speed' on matrices, joined forces with Born and Jordan to produce the 'three-man' paper *On Quantum Mechanics II*, received by *Zeitschrift für Physik* on 16 November 1925. Further progress was made in 1926, when Wolfgang Pauli independently calculated the wavelengths and intensities of lines of the Balmer series in hydrogen using matrix mechanics.

While this paper was under construction, Heisenberg received a long letter from Pauli, who wanted to know that if p and q do not commute, what would be the effect on simultaneous measurement of position and velocity. In Pauli's words:

> *One may view the world with a p-eye and one may view it with a q-eye, but if one opens both eyes at the same time, one goes crazy ...*[8]

Copenhagen and Uncertainty

Heisenberg spent the winter of 1926–1927 in Copenhagen. He and Bohr endlessly discussed how something so simple as the track of an electron in a cloud chamber can possibly be reconciled with the indeterminacy of the quantum mechanics which Heisenberg had created. In the end, Heisenberg was relieved when Bohr went skiing in Norway, giving him time to concentrate on the problem. One evening, after midnight, he suddenly remembered Einstein's words: '*It is the theory which describes what we can observe.*'[9] He went for a walk in a nearby park to think things over.

---

[8] Gino Segrè. *Faust in Copenhagen*. Pimlico, London. 2008.
[9] Werner Heisenberg. *Physics and Beyond*. Allen & Unwin, London. 1971.

In fact, he reflected, what we call the 'track' of the electron is a series of water droplets, all many times larger than the electron itself. He realized that the question to ask is:

Can quantum mechanics represent the fact that an electron finds itself approximately in a given place and that it moves approximately with a given velocity...?[7]

Heisenberg returned to the institute and was able to formulate what would become the uncertainty principle of quantum mechanics — the physical law that sets the limiting precision of simultaneous measurements of certain physical properties.

The commutation relation hypothesizes that the act of measurement alters the value of a physical property in an unpredictable way. Heisenberg had translated it from the mathematical language of matrices into a physical law.

More formally, Heisenberg's uncertainty principle states that:

Certain pairs of 'complementary' physical properties, such as momentum and position cannot be known with unlimited accuracy at the same time. The concept that exact values of these quantities exist together has no meaning.

The essence of quantum mechanics is contained in the uncertainty principle. It completely changes our concept of physical reality. The classical view is that physical objects have a 'real' existence with attributes which are independent of any observer. These attributes exist, and it does not matter how accurately we can measure them, or whether we can measure them at all. It is the 'common sense' view. But common sense does not apply in the quantum world. When we know the position of an electron we lose knowledge of its momentum, not because of our limitations in measuring ability, but because defined momentum does not exist together with defined position. For something like an electron, reality of existence takes on a different meaning.

It is not surprising that such a philosophy led to controversy and that fascinating arguments developed between Bohr, Einstein and other eminent protagonists. Einstein's immediate reaction is recorded in a letter to a colleague: '*Heisenberg has laid a large quantum egg. In Göttingen they believe it (I don't).*'

An Equation for Uncertainty

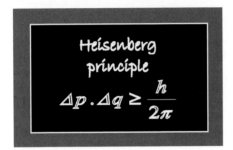

The uncertainty principle expresses the absolute limits of precision to which momentum and position can be known at the same time.

Uncertainty at Work — The Electron Forgets Where it was Going

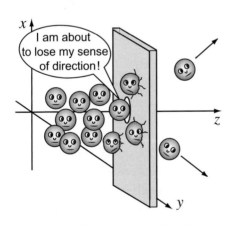

Electrons lose their sense of direction

The picture on the left is an 'artist's impression' of a real physical effect called *electron diffraction*. An electron beam is travelling from left to right, in the z direction. It comes to a barrier which stops all electrons except those which arrive exactly in the middle of a small opening. This is where the Heisenberg's uncertainty principle comes into play. Any electron which gets through has established its x and y coordinates by the act of passing through the opening. The more accurate is its defined position, the less well defined are the components of its momentum in the x and y directions i.e. 'sideways' to the beam direction. As a result these components are indeterminate and the transmitted electrons emerge at random angles, within limits determined by Heisenberg's principle.

The illustration is no more real than the picture of electron orbits in Bohr's atom, but the principle is true. It was tested experimentally less than two years later by passing a beam of electrons through a thin metal foil.[10]

The Nobel Prize

Heisenberg, Born and Jordan were nominated by Einstein for the 1932 Nobel Prize for physics. There was a delay in the announcement of the award and the ceremony did not take place until 1933, when the prize was given to Heisenberg. Heisenberg was embarrassed and immediately wrote to Born expressing his concern that he should be getting the award for work done in Göttingen by '*you, Jordan and I*'. He went on to say that their contribution to quantum mechanics cannot be changed by '*a wrong decision from the outside*'.

When Born was eventually honoured with the Nobel Prize in 1954, Heisenberg wrote an article in which he referred to contributions by Born and Jordan which, for 20 years, had not been '*adequately acknowledged in the public eye*'.

---

[10] See Chapter 9.

Chapter 9

Louis de Broglie — Matter Waves

De Broglie, the Student Prince

On 29 November 1924, Prince Louis de Broglie (1892–1987), a French noble-man, submitted his doctoral thesis to the Faculty of Science at the University of Paris. Such a thesis normally contains new experimental results and presents  conclusions, but does so within existing paradigms. This, however, was no ordinary thesis. It contained a strange new idea about the nature of matter and all material bodies. De Broglie put forward the hypothesis that matter, like light, exhibits both wave and particle characteristics. An electron, for example, behaves as a particle or as a wave, depending on the circumstances of the observation. The idea seemed quite absurd. Whatever about photons of light, which have no mass and are thus in a category of their own, conventional wisdom dictated that material particles have neither wavelength nor frequency, and they most certainly do not cancel one another and produce interference patterns.

As a rule, such radical ideas are unlikely to pass the rigorous questioning of a board of examiners. De Broglie's presentation was, however, well reasoned and logical. It was based on two main arguments:

1. Planck's quantum theory defines the energy of a light corpuscle in terms of the relation $E = hf$, an expression which contains

frequency. The concept of periodicity is therefore inherent in an entity which had been shown by Einstein, Compton and many others to have all the attributes of a particle.

2. The motions of electrons in an atom involve whole numbers. The only phenomena so far known in physics in which whole numbers are involved are resonant vibrations and interference, all of which are associated with waves.

Jean Perrin (1870–1942), the chair of the examination board, asked if there was any way in which experimental evidence of such *'material waves'* could be obtained. De Broglie's answer was that a stream of electrons passing through a sufficiently narrow opening should exhibit diffraction phenomena similar to the diffraction of light. The effect might be reproduced by sending electrons through a crystalline material.

From Conjecture to Fact

Neither de Broglie nor the members of the examination board were aware that the evidence was already in existence, hidden in the experimental results of two American physicists, Clinton J. Davisson (1881–1958) and Charles H. Kunsman (1890–1970) working at the Western Electric laboratory in New York City. They had bombarded a nickel crystal with electrons and had presented the reflected scattering data at the American Physical Society meeting in Chicago on 27 October 1921. More detailed accounts were to be found in two papers in *Physical Review* in 1923. The authors suggested that the way in which the electrons were reflected could be due to a process in which '*a fraction of the electrons penetrate the outer structure of the nickel atom and after executing simple orbits emerge without appreciable loss of energy*'. The question of wave-like behaviour did not arise. Nobody suspected that the pattern contained a vital clue to a mysterious property of all matter in the universe.

The only member of the examination board with expertise in quantum theory and relativity was the external examiner, Pierre Langevin (1872–1946). He sent a copy of de Broglie's thesis to

Einstein for a second opinion. Einstein replied with the words: '*I believe it is a first feeble ray of light on this one of our worst physics enigmas.*'[1] This was enough for the board of examiners; de Broglie was awarded his doctorate.

De Broglie's thesis was circulated in the academic domain and Walter Elsasser (1904–1991), an enterprising young research student at Göttingen, saw a copy in his university library. The researchers in Göttingen had discussed Davisson and Kunsman's experiments at departmental seminars and Elsasser wondered if there was a possible connection. Using the de Broglie formula for the wavelength of the incoming electrons, he made some calculations and found that the pattern in Davisson's data was consistent with the diffraction of electrons, resulting from de Broglie 'matter waves'.

Elsasser wrote a short note about his calculations and showed it to his research supervisor James Franck, who showed it to Born, the director of the institute. They both thought the idea was a bit crazy, but interesting enough to send it to *Naturwissenschaften* in Berlin. The editor didn't quite know what to do with it. Should he print such a radical suggestion by a 21-year-old student? He sought the advice of some eminent physicists who were just as unsure. Finally he showed it to Einstein, who happened to be in Berlin; Einstein is said to have replied: '*I think the man should be given a chance*'.[2] And so the note appeared, in July 1925, under the title '*Remarks on the quantum mechanics of free electrons*'. This was the first published recognition of experimental evidence that electrons were behaving like waves.

The Wavelength of Electron Waves is just as de Broglie had Predicted

At first Davisson did not believe Elsasser's conclusions, but soon changed his mind and started a new, more refined, series of

---

[1] Letter from Einstein to Lorentz. Einstein Archives. Hebrew University of Jerusalem. 16 December 1924.
[2] *Archive for the History of Quantum Physics.* 1962.

experiments, this time designed specifically to test the matter wave theory. Lester Germer (1896–1971) succeeded Kunsman as Davisson's assistant and by the end of 1927 they were able to publish unambiguous evidence of matter waves of wavelength exactly the same as those predicted by de Broglie's formula.

Simultaneously, in Aberdeen (Scotland), George Paget Thomson (1892–1975) and his graduate student Andrew Reid looked for matter waves, employing a slightly different technique. They also used a beam of electrons but, instead of reflecting them from a crystal, sent them through a thin metal foil. The transmitted electrons were recorded on a photographic plate behind the metal, producing a ring pattern.

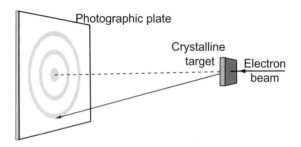

Electrons behave like waves

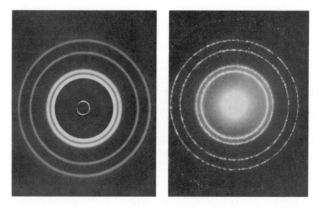

Diffraction patterns produced by X-rays (left) and electrons (right) on passing through aluminium foil

Thomson and Reid then repeated the experiment using X-rays with the same wavelength as the de Broglie wavelength of the electrons. By May 1927, they had obtained diffraction patterns in the form of concentric rings, as shown in the diagram. The two patterns are very similar, not only in their overall appearance, but also in those details which depend on the crystalline structure of the metal. De Broglie's theory had evolved from an unlikely hypothesis into an established fact.

To eliminate other possible explanations, Thomson carried out a number of tests. In one such test, each diffracted beam was directed through a magnetic field perpendicular to the original beam direction. This had no effect on the X-rays, but the electrons were deflected according to the laws of electromagnetism. This confirmed, beyond all doubt, that a beam of electrons had behaved as waves; the electrons had been diffracted by their passage through a metallic film. In Thomson's words: '*Once the particle appears the wave disappears like a dream when the sleeper awakens*'.[3]

It is extraordinary that John Joseph Thomson received the Nobel Prize in 1906 for the discovery of the electron, the first subatomic particle, and his son, George Paget Thomson (jointly with Clinton Davisson) received the Nobel Prize in 1937 for the discovery of electron waves.

De Broglie's Waves and Bohr's Quantum Orbits

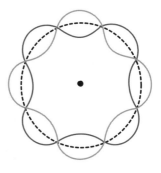

The wave fits neatly around the orbit

De Broglie re-interpreted the quantization of Bohr orbits in terms of standing waves. In the same way as a guitar string vibrates only at frequencies such that there are nodes at both ends of the string, the atom vibrates only at frequencies where the electron wave fits 'neatly' around the circumference of the orbit,

---

[3] G.P. Thomson. Nobel Speech. 1937.

forming a standing wave, as illustrated in the diagram. The result was mathematically equivalent to the quantization condition of Bohr's atom. So in a sense there was nothing 'new', apart from the idea!

The idea that there is some connection between particles and vibrations had previously been considered by a few unorthodox thinkers but never supported by proper scientific evidence. Baron Nicolai Dellingshausen published a work in 1872 entitled *Foundations of a Vibration Theory of Nature*, in which he identified atoms with resonant vibrations and described massive objects as '*extended centres of vibrational motions*'. More recently, in 1919, Marcel Brillouin had published a series of papers in *Comptes Rendus* on a vibrating model of an atom, to which de Broglie paid due credit in his thesis. It is unlikely, however, that de Broglie had ever heard of the work of Dellingshausen.

George Gamow (1904–1968) tells a nice story about de Broglie in his book *Thirty Years that Shook Physics*.[4] Apparently, Gamow decided to spend Christmas 1928 on holiday in Paris — his first visit to that city. He wrote to de Broglie asking if they could meet to discuss quantum theory and de Broglie suggested they meet at his home, as the university would be closed. De Broglie lived in a mansion in Neuilly-sur-Seine (a very fashionable suburb of Paris). When Gamow arrived, the door was opened by a '*very impressive*' butler. The conversation went as follows:

> Gamow: *Je veux voir Professeur de Broglie.* (I would like to see Professor de Broglie)
>
> Butler: *Vous voulez dire, Monsieur, le Duc de Broglie.* (If you please, Sir, Duke de Broglie)
>
> Gamow: '*O.K., le Duc de Broglie.*

---

[4] George Gamow. *Thirty Years That Shook Physics.* Doubleday & Co Ltd, New York. 1966.

Gamow was shown into a '*sumptuously furnished*' study, where he and de Broglie began to talk. Despite the fact that de Broglie spoke no English and Gamow's French was '*rather poor*', they managed to discuss matters of common interest. De Broglie was subsequently invited to give a lecture at the Royal Society of London. Gamow was in the audience and heard '*a brilliant lecture, in perfect English, with only a slight French accent.*' He concluded: '*Then I understood another of his principles: when foreigners come to France, they must speak French.*'

Chapter 10

Erwin Schrödinger —
Wave Mechanics

Schrödinger Enters the Scene

Born in Vienna on 12 August 1887, Erwin Schrödinger (1887–1961) was 14 years older than Heisenberg, with whom he was destined to play a central role in the quantum adventure. He entered the University of Vienna in 1906, graduating with a doctorate in 1910. His dissertation was on the conduction of electricity in moist air near the surfaces of insulators, a very practical topic, in contrast to the theoretical discoveries he was to make later in his career. In 1914, he received his *habilitation*[1] and almost immediately went to serve on the

Erwin Schrödinger. Courtesy of Austrian Post

Italian front in World War I. There he had the rare, if not unique, distinction of submitting and publishing a paper *The Acoustics of the Atmosphere* in *Physikalische Zeitschrift*, while at the battle zone in 1917. Evidently, scientific work did not interfere with his military duties, as he received a citation for outstanding service as battery commander during battle.

---

[1] 'Right of teaching' a specific academic subject at German universities.

While Schrödinger was in command of Austrian artillery firing at Italian troops, de Broglie was on the other side, in the wireless section of the French Army stationed at the Eiffel Tower in Paris.

At the end of the war, Schrödinger returned to Vienna and, after some movement to and fro between positions, as was quite common in German academic life at the time, he was appointed to the Chair of Theoretical Physics in Zürich in 1921. The six years spent in Zürich were the most active years of his academic life. He published his first paper on wave mechanics in 1926, within a year of the publication of Heisenberg's work on matrix mechanics. The 'new' quantum theory was largely developed by young men and sometimes dubbed '*knabenphysik*' — boy physics. Schrödinger was the exception; he was 41 years old when he made his first, momentous contribution.

A Complete Theory

From the moment Schrödinger heard of de Broglie's work, he was highly impressed. In December 1925, Einstein wrote to Schrödinger, saying that he had '*read the* [de Broglie's] *thesis with great interest in the ingenious theory*', which gave Schrödinger further encouragement to develop the wave concept into a complete theory of matter. The thesis dealt with 'free' electrons travelling through open space. A number of questions had to be addressed immediately: What happens to the waves of atomic electrons which, in Bohr's model, are in orbit around a nucleus? How far do the waves stretch out in space? Most importantly, how does the new theory merge with classical mechanics?

The 'old' mechanics had worked extremely well. It was simple and elegant, a 'scientific poem' based on solid foundations. Hamilton's principle of least action was a most reliable basis; it was unthinkable that it could be wrong. Now a '*strident disharmony*' had been introduced. A strange new condition had appeared, in the form of Planck's constant, which gave a numerical value for the minimum amount of action and restricted action to integral multiples of that value. A place had to be found for this *quantum* in the jigsaw of scientific knowledge. The problem was how to make it fit, without throwing everything else into disarray.

Schrödinger wanted to show that the quantum condition is not some strange add-on, but comes from the innermost essence of the theory. He argued that the new concept must be built on the same foundations as the well-established wave theory of light, where the principle of least time governs the path of light and is the basis of all optical phenomena. The wave nature of light is not apparent in large scale phenomena but becomes increasingly significant, and ultimately dominates, at small distances.

In Schrödinger's complete theory, the principle of least action also governs the path of particles and the behaviour of all physical systems. Schrödinger reasoned that, as with light, even though wave nature is inherent in all matter, the role of matter waves becomes significant only in the atomic world.

In his acceptance speech at the Nobel Prize ceremony on 12 December 1933, Schrödinger recalled his search for a unified concept of light and matter:

> *the analogy with the principle of the shortest travelling time of light is so close, that one was faced with a puzzle. It seemed that Nature had realized one and the same law twice by entirely different means: first in the case of light by the fairly obvious play on rays, and again, in the case of the mass points, which was anything but obvious, unless wave nature were to be attributed to them also.*

The Wave Function

The principle of Schrödinger's method is to describe the state of a particle or physical system by a mathematical expression called a *wave function*, so-called because it varies periodically like a wave or combination of waves. The wave function, usually designated by the symbol Ψ, can be a function of a particular coordinate (say, x) or, more generally, a function of space coordinates and of time.

The wave function is the de Broglie wave, expressed in a quantitative form and adjusted to fit a particular physical situation.

Schrödinger's method turned out to be much simpler and easier to visualize than Heisenberg's matrix mechanics, and was more readily accepted by most physicists. The puzzling thing was that the two methods of describing atomic behaviour appeared to be completely different.

The Wave Function of a Free Electron

In the simplest case of a free electron, propagating with momentum p, the wave function is a continuous sine wave of wavelength h/p, as given by the de Broglie formula.

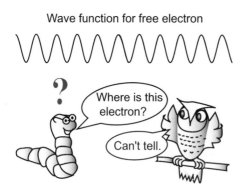

Wave function for free electron

Wave function of a free electron momentum h/p

The momentum of the electron is determined with a precision which depends on how well we can measure the wavelength. In this case, the wavelength, and hence the momentum, is known with unlimited precision. On the contrary, the wave amplitude is the same everywhere and the function tells us nothing about the position of the electron. Such a featureless wave represents a theoretical situation where we know everything about one physical variable, in this case momentum, but can say nothing about another observable, in this case position.

Superposing Momenta

A 'real life' electron is represented by a combination of waves superimposed to form the wave function. The wavelengths of these component sine waves vary over a range which corresponds to the intrinsic uncertainty in our knowledge of its momentum.

We can illustrate this in a diagram which shows a number of sine waves with slightly different wavelengths, together with the resultant wave which is the superposition of all of the sine waves. The amplitude of the resultant is largest where all the sine waves are in phase (at the centre of the diagram). As we go to the left and right of centre, the components are no longer exactly in phase and the resultant amplitude decreases. The resultant wave is called a *wave packet*.

In this *wave mechanical* representation of an electron, the electron is to be found somewhere in the wave packet, and the width of the wave packet represents the uncertainty in its position.

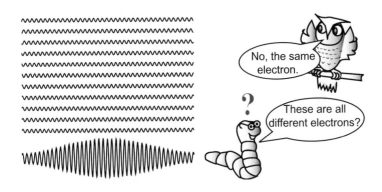

Wave packet — well-defined momentum

If the wavelengths of successive sine waves differ by only about 0.1%, the phase difference increases slowly as we move away from the centre and the wave packet is relatively broad, as is shown above. This means the position of the electron is not well determined. On the other hand, the wavelength is known to within a narrow range and therefore the momentum of the electron is well known.

If we now increase the difference between successive wavelengths by a factor of ten (to 1%), the phase difference increases much more rapidly as we move away from the centre and the wave

Wave packet — well-defined position

packet becomes narrower and sharper, as seen in the diagram above. Comparing the wave packets in the two diagrams, we can see that the position of the electron is now known more precisely, but at the expense of the momentum, which is less well defined due to the greater spread in wavelengths of the component waves.

All Roads Lead to the Uncertainty Principle

We can make only qualitative conclusions from these computer reconstructions, as the number of component waves and the range of wavelengths is arbitrary. A more precise mathematical relationship, known as the bandwith theorem, can be obtained using Fourier techniques. It can be written in the form $\Delta x \cdot \Delta k = h/2\pi$ and gives the relation between the width of the wave packet Δx and the range of wavelengths[2] of the components. (The formula is often used in radio communication and signal processing.)

If we use de Broglie's formula for wavelength, $\lambda = h/p$, we arrive at Heisenberg's uncertainty relationship:

$$\Delta p \cdot \Delta x = h/2\pi.$$

Going by the apparently quite different routes of Schrödinger's waves and Heisenberg's matrices, we arrive at the uncertainty principle.

The Superposition of States and the Collapse of the Wave Function

Every wave tells a story.

The wave diagrams tell us about two fundamental features of wave mechanics. The wave packet which represents a particle, in this case an electron, tells us something about its position and something about its momentum. However the information is somewhat clouded. The particle is in a *superposition of states* of position and of momentum. The instant we make a measurement of one of these

---

[2] k is the inverse wavelength $k = 1/\lambda$.

observables, the wave function undergoes what is sometimes rather dramatically called a *collapse* and all knowledge of the other observable 'evaporates'. If we measure the momentum of the particle, we obtain a single sine wave of appropriate wavelength. Conversely, if we measure its position, the wave function collapses into an infinitesimally thin wave packet at the measured value of x.

The Schrödinger Equations

Any wave, whether sound, light, or the wave on a guitar string is described by an equation. There was no equation to describe de Broglie's waves; he had not attempted to derive one. Schrödinger set himself the task of finding the wave equation for de Broglie waves. At a colloquium in Zürich towards the end of November 1925, Schrödinger gave a 'beautifully clear account' of de Broglie's work, but not everyone could accept the new concept. Peter Debye (1884–1966), head of the research group in Zürich, even going so far as to describe the work as: '*rather childish*'.[3]

Schrödinger continued to work on the development of a wave equation and, as Christmas approached, he decided to take a two-week holiday in Arosa, an alpine resort that he was especially fond of. Unlike Heisenberg in Heligoland, Schrödinger was by no means free from distractions but, nevertheless, was inspired to great things and published the first of his wave equations shortly after returning home.

The *Schrödinger equations* are modelled on the classical Hamilton equations of motion. They represent the conditions the wave function must satisfy, so as to conform to the laws of nature.

[3] Felix Bloch. *Reminiscences of Heisenberg and the Early Days of Quantum Mechanics.* Physics Today 29, 23–27 (1976).

Classical mechanics was first summarized by Newton in two statements and one equation. These were later put in a more general form and derived from the principle of least action by Hamilton. Heisenberg adapted Planck's quantum condition and expressed them in the new mathematical language of matrix mechanics, and now Schrödinger translated them into yet another language, that of wave mechanics.

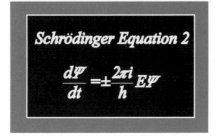

Schrödinger's first equation was published in January 1926 in *Annalen der Physik* and is known as the *time independent Schrödinger equation*. It is a statement of conservation of energy, translated into 'Schrödinger language'. As the name implies, it specifies a condition that a wave function must fulfill at all times.

Schrödinger's second equation describes how the wave function changes with time; it is the wave-mechanical equivalent of Hamilton's equation of motion.

The significance of Schrödinger's work is described by Richard Feynman in these words:

the great historical moment marking the quantum mechanical description of matter occurred when Schrödinger wrote down his equation in 1926. Where did we get that from? Nowhere. It's not possible to derive it from anything you know. It came out of the mind of Schrödinger.[4]

---

[4] Richard Feynman. *Feynman Lectures on Physics.* Vol. 3, 16.4 and 16.12. Addison Wesley, Reading, Mass. 1965.

Potential Well

Wave mechanics is modelled on the generalized mechanics of Lagrange and Hamilton, in which force is represented by potential energy. The *force field*[5] is represented by a *potential well.*

Using a little artistic license, let us talk about an electron which accidentally stumbles into such a well. Suddenly the ground underneath him gives way and he falls in. Classically, we would say that he is pulled in by a force. In wave mechanics, the nature of that force is described by the size and shape of the well, while the energy required to pull him out depends on the depth of the well.

The simplest, if somewhat idealized, example is that of a particle such as an electron in a *one-dimensional potential well of infinite depth*, often called a 'particle in a box'. Infinite depth means that an infinite amount of energy would be required to liberate the particle; the walls of this box are impenetrable. We might visualize the particle bouncing around between two thick, high walls. It has a lot of kinetic energy but not enough to clear the wall or tunnel through it. We must stretch our imagi-

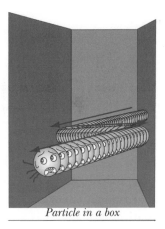

Particle in a box

nation a little further; there is no friction and the ball is perfectly elastic. Classically, there are no restrictions on its movement *within* the well; it can be found anywhere inside the box at a given moment. This is not so for the 'imprisoned' electron.

---

[5] A force field describes how a force behaves at different points in space.

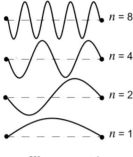

$n = 8$

$n = 4$

$n = 2$

$n = 1$

Waves on a string

In Schrödinger's model, the wave function must fit into the well in such a way that there is a node at each boundary,[6] so the probability of finding the particle outside the well is zero. The wave function can have only certain wavelengths, like the characteristic wavelengths of a vibrating guitar string. The mathematics is the same in both cases; just as only certain vibrational frequencies exist in a guitar string, so the particle is restricted to certain energy levels.

There is, however, a very interesting difference; the stretched string need not vibrate at all; it can have zero energy but the lowest energy level of the particle corresponds to $n = 1$, the fundamental mode in the diagram. The electron cannot have zero energy, it cannot stand still!

The Hydrogen Atom

The nucleus of the hydrogen atom attracts the electron with a force described by Coulomb's inverse square law. The potential well created by the electric charge of the nucleus is the classical Coulomb potential, which varies inversely as the separation. In contrast to the artificial one-dimensional square well, this is the 'real world'; there are three dimensions, the potential is more difficult to visualize and the wave function is much harder to handle mathematically. It is a measure of Schrödinger's work rate that he solved the problem of the hydrogen atom in a matter of months after his original paper.

---

[6] If we want to be pedantic, the node is at an infinitesimal distance outside the boundary.

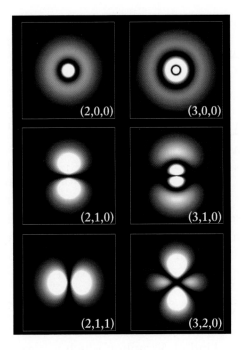

Hydrogen wave functions.
Courtesy of Wolfgang Christian, Davidson College, North Carolina

Some two-dimensional computer-generated hydrogen wave functions are shown in the diagram above.[7] The electron is most likely to be found in the white areas and least likely to be found in the purple areas.

Two Roads to the Same Goal

In May 1926, Schrödinger published what was probably his most significant paper, in which he showed that his wave mechanics and Heisenberg's matrix mechanics are equivalent. The representation may be different but the mathematical basis is the same. In each case, the theory leads to the quantum structure of nature as a logical consequence.

---

[7] The set of quantum numbers associated with each wave function is shown at the bottom right of the corresponding picture.

Interpretation of the Wave Function

While the mathematical methods appeared to be falling into place, the philosophical meaning of quantum mechanics was becoming more mysterious. What is the nature of the Schrödinger waves?

Waves in liquids, gases and even solids are very real and easy to visualize. They require a medium and involve a transfer of energy. Light and other electromagnetic waves do not require a medium. They involve a transfer of a physical condition which is a combination of electric and magnetic fields. The resultant disturbance can be detected by the human eye or a radio antenna, not to mention physical effects such as the emission of photoelectrons.

While the wave function could be visualized and plotted in graphical form, Schrödinger was not sure at first what it meant. Does it represent some kind of disruption which is travelling in wave-like form? Does it carry energy? What is the significance of the amplitude of the wave?

Probability

Max Born was the first to suggest that Schrödinger's wave function describes 'waves of probability'. The amplitude, or more precisely, the *square of the amplitude* represents the *probability density* of finding the particle in a certain place at a certain time. The abstract concept of 'likelihood' is represented and visualized by the mathematics of the wave function.

The philosophical conundrum presented by physical laws based on probability caused Schrödinger to have second thoughts about the basis of the quantum mechanics he had just invented.

Schrödinger wasn't the only one with major misgivings. Einstein, who had contributed so much at the beginning, when he introduced Planck's hypothesis into the theory of the photoelectric effect, could not come to terms with the probabilistic aspects of quantum mechanics. He firmly believed that '*God does not play dice*' and the theory was '*an incomplete description of reality*'.

Bohr's attitude was quite different. He described his awe of the new theory he himself was helping to create with the words: '*If quantum mechanics hasn't profoundly shocked you, you haven't understood it*'.

Einstein did not believe in a universe governed by chance

Schrödinger Tries to Avoid 'Probabilities'

In July 1926, Schrödinger was invited to the Institute for Experimental Physics in Munich to give two lectures on wave mechanics. Using the hydrogen atom as an illustration, he showed how wave mechanics gave a simple and elegant solution to a problem which was very difficult to solve using quantum mechanics. His audience was enthusiastic; maybe there was no need to learn the strange mathematics of matrices after all?

Schrödinger then went on to discuss the wider implications of wave mechanics, as he saw them. His view was that everything comes down to waves; particles are just tightly bunched wave packets that give the illusion of being discrete. There are no quantum jumps, just smooth transitions from one standing wave to another.

Werner Heisenberg was at both lectures. He was enthusiastic, but only to the point where Schrödinger began to talk about his own interpretation of the wave function. The discussion after the second lecture took a dramatic turn. Heisenberg pointed out that Schrödinger's interpretation would not even explain Planck's radiation law.

Heisenberg's remarks were, to say the very least, unwelcome. Wilhelm Wien, the director of the institute, was in the audience and stepped in, effectively taking Schrödinger's side and saying that he understood Heisenberg's disappointment that quantum mechanics was finished; any outstanding difficulties with wave mechanics would surely be resolved by Schrödinger in the near future.

Bohr Invites Schrödinger to Copenhagen

When Bohr heard what had happened, he invited Schrödinger to his house in Copenhagen so that they would have an opportunity to discuss Schrödinger's ideas in detail. From the time they met at the railway station, the two men spent their days in intense and relentless discussion, on which Bohr thrived.

Heisenberg who had returned to Copenhagen to 'observe the proceedings' gives the flavour of the discussions in his book, *Physics and Beyond*:[8]

Schrödinger: ... then we must ask ourselves how precisely the electron behaves during the jump ... what laws govern its motion during the jump? In other words, the whole idea of quantum jumps is sheer fantasy.

Bohr: What you say is absolutely correct. But it does not prove that there are no quantum jumps. It only proves that we cannot imagine them, that the representational concepts with which we describe events in daily life ... are inadequate when it come to describing quantum jumps.

Schrödinger: If all this damned quantum jumping were really here to stay, I should be sorry I ever got involved with quantum theory.

Bohr: But the rest of us are extremely grateful that you did: your wave mechanics has contributed so much to mathematical clarity and simplicity

---

[8] Werner Heisenberg. *Physics and Beyond: Encounters and Conversations.* Allen & Unwin, London. 1971.

that it represents a gigantic advance over all previous forms of quantum mechanics.

After a few days, Schrödinger became ill and had to stay in bed. Even then, while Mrs Bohr was nursing him and feeding him tea and cake, Niels sat on the edge of the bed and continued talking at Schrödinger.

Chapter 11

Eigenstates — The Theory of the Seen and Unseen

Eigenstates

The German word *der Eigenwert*, (literally, own worth) is a mathematical term which occurs in both matrix mechanics and wave mechanics. As is so often the case, the German language expresses, in one word, a concept which other languages may take a whole sentence to describe. The anglicized version, *eigenvalue*, is defined in quantum mechanics as the characteristic or intrinsic value of a physical observable or system.

The process of measurement implies something in quantum mechanics which is so foreign to the household world that we have to use new words, in this case borrowed from mathematics, to describe simple concepts like the value of a physical quantity. In classical physics, when we measure something, we take it for granted that the information is already there and exists independently of the observer. In practice, the measuring process may disturb the system but that is irrelevant to the principle. For example, it may be difficult to measure a soap bubble without bursting it. Nevertheless, it goes without saying that the existence and properties of the bubble are independent of whether or not it we are aware of it. While it may have been unchallenged in the classical household world,[1] that very question was central to what became known as the *Copenhagen interpretation* of quantum mechanics.

---

[1] An exception must be noted in the philosophy of *Immaterialism* proposed in the 18th century by George Berkeley, as we shall see later.

A fundamental feature of quantum mechanics is that before a physical system is observed and measured, it exists in a mysterious state of *quantum reality*. In a mathematical formalism it is described in terms of a *superposition of possible states*, known as *eigenstates*, of a particular physical variable.

We saw in the previous chapter that the wave function of a free electron of known momentum is a continuous sine wave with a precisely defined wavelength. An exact measurement of another observable, in this case its position, would cause this function to collapse into an infinitely thin wave packet. The electron has 'jumped' from an eigenstate of momentum into an eigenstate of position. (A free electron has an infinite number of eigenstates, of both momentum and of position; it is not constrained.)

Another example is the hydrogen atom, which has a series of energy eigenstates, corresponding to the energy levels in the Bohr model of the atom. The act of observing an eigenstate causes the system to assume that state. Only after it is observed does the state become a physical reality. It comes as no surprise that this is one of the aspects of quantum mechanics which became a 'bone of contention' in the early days of the quantum adventure. Schrödinger, who had played such a central part in its development, became very disenchanted by implications of the very theory which he himself had helped to create.

Schrödinger's Cat Paradox

In 1935, Schrödinger devised a thought experiment to show that the implications of quantum theory are absurd.[2] It is known as the paradox of Schrödinger's cat.

The cat, unseen and seen

In Schrödinger's experiment, a live cat is put into a box containing a small amount of radioactive material, a Geiger counter and a bottle

---

[2] E. Schrödinger. *Die gegenwärtige Situation in der Quantenmechanik.* Naturwissenschaften 23, 807–812 (1935).

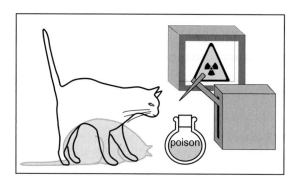

Schrödinger's cat

of poison. The experiment is set up so that there is a 50–50 chance of a radioactive decay during the time the counter is switched on. If an atom decays, the counter discharges, triggering the release of a hammer which breaks the bottle. The poison is released and the cat dies. If not, the cat lives.

According to the laws of the household world, there is a 50–50 chance of finding the cat alive when we open the box. More to the point, even before we open the box we take it for granted[3] that the cat will be in one of two states i.e. alive or dead.

Quantum mechanics presents a very different and quite bizarre picture. It asserts that, until we make a measurement (open the box, or even peep through a tiny window), the radioactive atom is in a sort of limbo, it is neither 'whole' nor decayed. The entire apparatus, including the cat, is described by the superposition of two quantum states — one in which the cat is alive and the other in which it is dead.

In Schrödinger's words:

> *The [wavefunction] of the entire system would express this by having in it the living and dead cat (pardon the expression) mixed or smeared out in equal parts.*

Only when we open the box does the wave function collapse to one of the two states; we see either a live cat or a dead cat.

---

[3] Unless we follow Berkeley philosophy of *Immaterialism.*

The household world does not exhibit the weird and counter-intuitive behaviour of the atomic world. A cat may have nine lives, but whoever heard of a cat at the same time alive and dead?

Absurd as it may be in the case of a cat, simultaneous existence in a number of 'virtual' states forms the essence of quantum mechanics. As will be described below, a simple table-top experiment demonstrates photons with characteristics arguably just as illogical as those of Schrödinger's example. The experiment involves light which is *polarized.*

Polarization

Double refraction.
Courtesy of Adrian Pingstone

In 1669, Erasmus Bartholin (1625–1698), a Danish mathematician at the University of Copenhagen, discovered that if he placed a crystal of calcite (crystallized calcium carbonate) on top of a drawing or writing, he would see two images instead of one. Christiaan Huygens (1629–1695) investigates this *'strange refraction of light of a particular crystal brought from Iceland'* in his work *Traité de la Lumière*, published in 1690. Huygens believed that light was a wave and had developed a geometrical theory of light propagation. Based on this, he concluded that the splitting was in some way due to the fact that the speed of the light depends on the direction in which the light passes through the material.

In 1809, Étienne Louis Malus (1775–1812) examined reflected sunlight through a calcite crystal. As he turned the crystal the two images of the sun became weaker and stronger alternately. Clearly, there was something different about the two rays, apart from their speeds:

> *one evening he [Malus] was looking through a calcite crystal at the reflection of the setting sun in a window of the Luxembourg palace and was*

astonished to see for certain orientations only one image instead of two, the ordinary and extraordinary images appearing and disappearing in turn on rotating the crystal.[4]

The Classical Model of Polarization

Thomas Young (1773–1829), who had demonstrated the interference of light six years earlier, was very interested in Malus' work. Young and, independently, Augustin Fresnel (1788–1827) produced a mathematical analysis of *polarization* in 1817. Their analysis was based on the following hypothesis:

Light waves vibrate in all directions in a plane perpendicular to the direction of propagation. When light is *polarized,* the vibrations are restricted to a single direction. A crystal such as calcite has an *optical axis* and splits the incoming unpolarized light into two rays, one polarized parallel to the optical axis and the other perpendicular, as illustrated in the diagram.

One of these rays, the *ordinary* ray travels through the crystal with the same speed in all directions; it has a single index of refraction and behaves like a normal ray of light. The *extraordinary* ray does

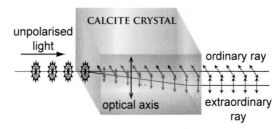

Polarization by double defraction

not obey Snell's law of refraction because its speed inside the crystal depends on the direction of propagation.

Nowadays, polarization can easily be achieved with synthetic materials, commercially known as *polarizers*. These are produced in thin sheets and widely applied to reduce glare (in sunglasses and car

---

[4] Ernst Mach. *The Principles of Physical Optics.* Dover Phoenix Editions, New York. 2003.

windscreens, for example). Ideally, in these materials, the perpendicularly polarized light component is absorbed and the component polarized parallel to the optical axis is transmitted, although they are never 100% efficient.

Polarized light is also routinely produced in the natural environment, whenever light is reflected.

Crossed Polarizers

Polarization may be demonstrated by using two polarizers. The incoming light is polarized by the first polarizer and 'analysed' by the second.

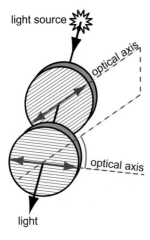

Photon passes through two polarizers

If the optical axes of the two polarizers are parallel, all the light is transmitted; if the axes are at right angles (*crossed* polarizers) all the light is absorbed.

To see what happens between these two extremes, we can turn one polarizer slowly so that the axes are no longer parallel. We find that light is still transmitted, but its intensity is reduced by a factor of $(\cos \theta)^2$, where θ is the angle between the axes, as illustrated in the diagram. This experimental result was discovered by Malus, who was awarded a prize for his work in 1810 by the Académie des Sciences in Paris. Surprisingly, one year later, he was awarded the Rumford Medal of the Royal Society of London, even though France and England were at war at the time.

Inserting a Third Polarizer

We start with two polarizers adjusted so as to be perfectly crossed, and then introduce a third polarizer. We will hold it at various random angles, so that its optic axis can point in any direction.

When we place this additional polar-izer in front of or behind the combina-tion, there is no change. The crossed polarizers block the light completely and the third filter makes no difference.

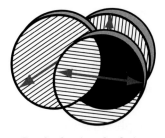

However, something surprising does happen when we put the randomly orien-ted polarizer between the two crossed polarizers. Suddenly light gets though!

Randomly oriented polarizer between two crossed polarizers

By inserting another polarizer into the path of the wave, we have given it a chance to pass through the system. Paradoxically, the new obstacle has made the task easier and not harder.

Can Photons be Polarized?

This brings us to the 'million dollar' question: 'Is quantum theory inconsistent with polarization? Can *photons* be polarized?'

The quantum mechani-cal description of polariza-tion is quite different from the classical wave representa-tion. There is no such thing as a fraction of a photon so when a photon comes to a polarizer, it is either transmit-

ted or absorbed; there are no half measures! An experiment can give just one of two possible answers: YES or NO. The photon faces a test which it either passes or fails.

If and when a photon passes the test, it takes on a definite identity. In quantum mechanical language it jumps into an *eigen-state of polarization in a certain direction*. If, as it continues on its way, it comes to a second polarizer with an optical axis oriented in exactly the same direction as the first, it will certainly pass; simi-larly if the axis is orientated in a perpendicular direction, it will certainly fail. For any other orientation we cannot predict whether the photon will pass or fail, all we know is that the statistical probability of passing such a second examination is proportional to $(\cos \theta)^2$.

The important feature of this representation is that once a photon, or any other quantum mechanical system, is in a certain eigenstate, its past history has been wiped out. It has 'forgotten' any previous tests. All the photons which pass through a particular polarizer are in the same state of polarization when they emerge, regardless of what happened to them beforehand.

Photons Travelling Through Multiple Polarizers

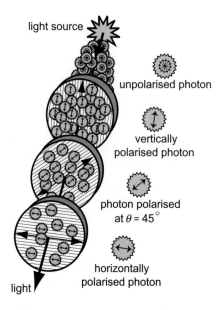

light source

unpolarised photon

vertically polarised photon

photon polarised at $\theta = 45°$

horizontally polarised photon

light

Photon passes through a series of polarizers

The diagram gives a schematic illustration of an experiment in which a photon passes through a series of polarizers. Every photon leaving the light source is 'unobserved' as regards polarization. It is in a *superposition of states of polarization.* Since the polarization vector can point in any direction, there is an infinite number of these states. Quantum mechanically we describe it as a two-state system of polarization; parallel and perpendicular to a particular direction in space. This direction is determined by the optical axis of the polarizer.

The first polarizer determines whether the photon is in the vertical or horizontal state of polarization and lets it through only if it is vertically polarized.

The next polarizer defines a new direction at 45° to the first. Since $(\cos 45°)^2 = 0.5$, each photon has a 50% chance of getting through. If the photon does get through, it promptly forgets it was ever vertically polarized. It now has a 50% chance of getting through the final, horizontal polarizer.

The Mysterious Logic of the Quantum World

The following analogy of the polarization of photons has been taken from the book *Quantum Reality* by Nick Herbert[5]; it is attributed to the philosopher of science, Ariadna Chernavska:

> *Suppose we pass cattle through a gate which only lets through horses and rejects all cows. Next we pass these horses through a second gate which lets through only grey animals and rejects all black ones. Only animals which are both horses and grey can pass through both gates. To our surprise, approximately half of such animals turn out to be cows!*

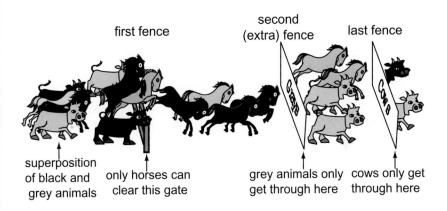

The first fence determines whether the animal is a horse. The second fence determines colour. The animals which get through *this* fence are grey, but they have lost their identity as horses or cows. They are a 50–50 mixture of grey horses and grey cows. Without the fence in the middle, the combination of the first and last fence would block all animals.

---

[5] Nick Herbert. *Quantum Reality. Beyond the New Physics.* Anchor Press/Doubleday, New York. 1985.

Are We Making it all too Complicated?

Has all this been made more complicated than it really is? Could we ignore the fact that light consists of photons and just use the classical wave model, which includes polarization?

We could say that the plane of polarization is changed, at the expense of a loss of amplitude, when the photon passes through a polarizer. We would need some ingenuity to explain the results of the experiments we have just discussed; would that explanation be more credible than this business with eigenstates?

The answer came in 1960, when an experiment unambiguously demonstrated the same sort of bizarre quantum behaviour by a subnuclear particle called the kaon. To describe this quantum behavior and what it means, we must first enter the world of *elementary particles* and introduce the kaon and some of its 'peers', which we will do in the next chapter.

Chapter 12

Eigenstates in the Subnuclear World

To get first-hand experience of quantum mechanics in action, we must go to the world of elementary particles where quantum laws are supreme and mixtures of eigenstates are commonplace. The existence of another subnuclear particle (in addition to the proton, neutron and electron) was predicted by Hideki Yukawa (1907–1981) at a meeting of the Physico-Mathematical Society in Osaka, Japan in 1934.

Yukawa used a clever application of the methods of quantum
mechanics, in which the *nucleon* is
considered as a two-state system
with the proton and neutron as
eigenstates. He even determined
the mass of the new particle to be
about 200 times the mass of the
electron. Communication between
countries such as Japan, Europe
and America was not as good as it is
today and his paper did not attract
the publicity it deserved. World War
II soon intervened and there was no
time or inclination to look for the
particle.

*Hideki Yukawa. Courtesy
of Japan Post*

When the war was over it was realized that the Yukawa particle, if it existed, was key to the understanding of nuclear forces. There was no question of observing it directly, just as it is impossible to 'see' an electron or a proton. It is not even a constituent of ordinary matter and would most probably make only a fleeting appearance for perhaps one millionth part of a second after it was created.

According to Einstein's theory of relativity, it would be possible to create such particles, provided a large amount of energy was concentrated in a tiny volume of nuclear dimensions. At that time there was only one possible source of such energy — and these were particles known as cosmic rays, which were known to bombard the Earth's atmosphere.

Cosmic Radiation

The discovery of cosmic rays is usually credited to Victor Hess (1883–1964) who in 1912, at considerable danger to himself, undertook a series of balloon flights at high altitudes to measure radiation levels. He found that the radiation level was several times greater at an altitude of five km than at sea level and concluded:

> *this ionization might be attributed to the penetration of the Earth's atmosphere from outer space by hitherto unknown radiation of exceptionally high penetrating capacity, which was still able to ionize the air at the Earth's surface noticeably.*[1]

Detection of Cosmic Rays

In the ensuing years, various detectors were developed to investigate the nature of cosmic radiation. Of particular interest is the detector built in 1911 by C.T.R. Wilson.

Charles Thomson Rees Wilson (1869–1959) graduated from Sidney Sussex College, Cambridge in 1892. Interested in meteorology, he spent a few weeks in September 1894 as a meteorological observer investigating weather conditions on Ben Nevis. He was particularly impressed by the effects of sunlight on the clouds around the summit of the mountain and decided to try and reproduce them in the laboratory.

---

[1] Victor Hess. Nobel lecture 1936.

Cloud Chamber

Wilson built a metal chamber and filled it with air which was saturated with water vapour. The design of the chamber allowed for sudden expansions which would cool the air, causing condensation of water droplets and cloud formation. Wilson could now make the clouds 'to order' and study them in the laboratory.

C.T.R. Wilson

He illuminated the chamber with flashes of light and photographed the formation of water droplets around dust particles in the air. Soon he discovered something which promised to be of much more interest than the optical phenomena which he had intended to study. Under certain conditions, the droplets would form in a completely dust-free environment. What was acting as the seed for their formation? Wilson speculated:

> *(This observation)... at once suggested that we had a means of making visible and counting certain individual molecules or atoms which were at the moment in some exceptional condition. Could they be electrically charged atoms or ions?*[2]

He was right. His cloud chamber turned out to be a window for observing effects caused by individual charged atoms (*ions*). The atoms themselves are not observed but, like a snowball which creates an avalanche, they act as seeds for droplets much larger than themselves.

As the charged particles move past normal atoms they disturb orbital electrons, triggering the formation of droplets and providing

---

[2] C.T.R Wilson. Nobel lecture. 1927.

a record of their movements. By 1911, Wilson had developed his method and could photograph tracks of alpha particles (positively charged helium nuclei) and beta particles (electrons).

Subsequently, the cloud chamber was improved and developed by Wilson himself and by P.M.S. Blackett (1897–1974) who, in the early 1930s designed a 'triggered' cloud chamber which was much more efficient because the cosmic rays themselves caused the chamber to expand and, in that sense, could 'take their own photographs'. The photographs were taken within 1/100th of a second of a cosmic ray passing through the chamber, while its own track and those of any product particles were still 'fresh'.

As soon as World War II ended, laboratories were set up both at ground level and at high mountain altitudes to look for the Yukawa particle, or any other products of cosmic ray interactions.

An Unexpected Particle Makes its Appearance

On 15 October 1946, at Manchester University, George Rochester (1908–2001) and Clifford Butler (1922–1999) noticed something unusual in one of their cloud chamber photographs.

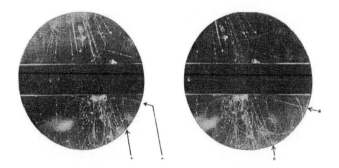

Cloud chamber photographs of V particles

Two pictures, taken from different angles, showed a pair of tracks which appeared to diverge from a single point, just below a lead plate at the centre of the chamber. Analysis of these stereoscopic views confirmed that what looked like a 'V' was not due to the accidental superposition of two separate tracks but was due to two individual particles which appeared, as if from nowhere. Rochester and Butler suggested that what they were seeing was the decay of an electrically neutral particle, created in a cosmic ray interaction in the lead plate. Uncharged particles do not leave tracks in a cloud chamber so the particle itself would not have been visible. Fortunately, it turned out to be highly unstable and decayed spontaneously, a short distance below the plate, into two lighter particles of equal and opposite charge. The 'V' shows those particles leaving the scene.

If the suggested explanation was true, this was a new subnuclear particle. It was first called a 'V particle' but later became known as the *k-meson or kaon*, derived from the Greek word μεσον meaning *middle*, because the mass of the kaon is somewhere between the masses of the electron and the proton.

No further 'V events' were seen at ground level for two years; Rochester described this as: '*tantalizing and embarrassing for the Manchester group*'. Then, in 1949, good news came from Pasedena in California where Carl Anderson (1905–1991) was doing cloud chamber experiments at high altitude:

> *Rochester and Butler may be glad to hear that we have 30 cases of forked tracks similar to those they described in their article in Nature about two years ago, and so far as we can see now their interpretation of these events as caused by new unstable particles seems to be borne out in our experiments.*

New Particles Made to Order

In the early 1950s, powerful proton accelerators were built and it was no longer necessary to expose detectors to cosmic radiation, in the hope that something interesting might happen. Protons were directed into a tungsten target, initiating very large numbers of nuclear interactions under controlled conditions. The energy of the accelerated protons was converted into matter at the rate of exchange, $E = mc^2$, predicted by Einstein over 50 years earlier. The matter appeared in the form of myriads of subnuclear particles, some known, others which had never been seen before.

The existence of the particles seen in cosmic ray interactions was confirmed, together with large numbers of new elementary particles. Some were more and others less massive than the proton. The mass of the kaon was measured more accurately and found to be slightly more than half (0.526) the mass of the proton.

All the new particles had one thing in common — they were unstable and decayed into other, lighter particles within tiny fractions of a second.

Quantum Numbers

Physicists were finding so many particles that they were hard-pressed to keep track of them, much less to work out a decent theory to explain what was going on. To create some sort of order in the 'particle zoo', particles with similar properties were assigned to specific groups. A simple and efficient way of doing this is to label the particles with *quantum numbers* which reflect their behaviour. To restrict the interactions between particles to those which were observed (or, more importantly, to give reasons why certain perfectly plausible interactions were not observed), ad hoc rules were introduced.

Conservation Laws

The ad hoc rules came in the form of new *conservation laws*. Well-established laws such as the conservation of energy and conservation of electric charge always apply, even in the weird quantum world. Charge comes in integer multiples, or quanta, of the charge on an electron ($\pm e$) and the sum of these quanta must remain the same before and after any sort of interaction. The particles in the zoo interact in different ways. At one extreme the *strong interaction*, which binds protons and neutrons together in the nucleus, breaks very few rules. It takes place extremely rapidly (in about 10^{-23} seconds, which is approximately equal to the time taken by light to go from one side of a proton to the other). We can never see direct evidence of unstable particles which decay via the strong interaction, it all happens too fast. On the other hand, the so-called *weak interaction* breaks the most rules. Many conservation laws are violated in weak interactions, which take place in about one billionth part of a second, very long on a nuclear scale. Radioactive decay is the result of a weak interaction.

The Strange Nature of the Kaon

It soon became apparent that the kaon did not behave like any other particle known at that time. It is produced in large numbers, typical of the strong interaction, but decays comparatively slowly. This means that, although it is created in a strong interaction, it is unable to decay in the same manner. Murray Gell-Mann (born in 1929) explained this in 1953 by inventing a new quantum number, which he facetiously called *strangeness*; this new quantum number, like many others, is conserved in the strong interaction but not in the decay which takes place via a weak interaction.

The positively charged kaon was arbitrarily assigned a strangeness $S = +1$ and the negatively charged kaon a strangeness of $S = -1$. The name 'strangeness' was well chosen as it happens; another member of the kaon family would be seen to exhibit properties that were even stranger, very much characteristic of the quantum world!

The Neutral Kaon Demonstrates Mysteries of the Quantum World

The kaon first observed by Rochester and Butler was electrically neutral and the one which, as it turned out, exhibited the most puzzling behaviour. It seemed to have a dual personality; sometimes it behaved as if it had strangeness of +1 and at other times as if it had strangeness of −1. In addition, its lifetime could be either 'short' (about one ten-billionth of a second) or 'long' — about 600 times longer — but with nothing in between.

There was no apparent correlation between strangeness and lifetime, so it was unlikely that there were two different particles. The simplest explanation seemed to be that the neutral kaon was a single particle behaving like Robert Louis Stevenson's Dr Jekyll and Mr Hyde, showing the two sides of its character in an unpredictable manner.

The Gell-Mann–Pais Scheme

In 1955, Murray Gell-Mann and Abraham Pais (1918–2000) realized that this maverick particle could be viewed as a superposition of quantum states. Applying quantum mechanical rules and pure logic, they created a beautiful model in which the kaon's 'strange' properties fell into place. The model also produced a number of new predictions and, with them, a challenge for experimentalists.

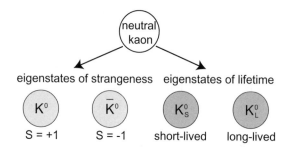

Eigenstates of the neutral kaon

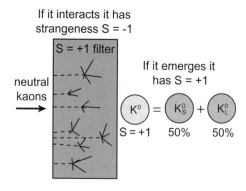

Selecting neutral kaons with strangeness +1

According to Gell-Mann and Pais the neutral kaon is a two state system of strangeness. If it is in the $S = -1$ state, it is likely to be absorbed by interactions which create other $S = -1$ particles. If it is in the $S = +1$ state, it has no such modes of interaction. If a neutral kaon beam passes through a block of material, the S = −1 component will be absorbed, while the S = +1 component will pass through. The block of matter is equivalent to a detector which measures strangeness.

According to the rules of quantum mechanics, a kaon which has passed through the block is in an eigenstate S = +1. At the same time it is a 50–50 mixture of states of lifetime. Its life expectancy is either 10^{-10} s or 600 times longer.

The first moments are critical!

Regeneration of Short-Lived Kaons

Soon after publishing his paper with Gell-Mann, Pais co-authored another paper, this time with Oreste Piccioni (1915–2002), on *How to verify experimentally a recent theoretical suggestion that the K^0 meson is a particle mixture*. Their suggestion was to make two measurements in succession:

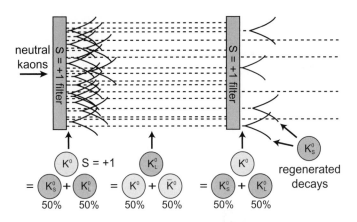

How to demonstrate that the neutral kaon is a 'particle mixture'

In the first part of the experiment, a beam of S = +1 neutral kaons is produced using a filter, as we have already described. The second part of the experiment is passive and involves waiting for at least ten times the mean life of the short-lived K_s^0 component. During this time, the neutral kaons travel to a second identical filter, about 4 m away. By the time they arrive at that filter, 99.99% of the short-lived particles will have been eliminated, leaving an almost pure beam of long-lived K_L^0 which is a 50–50 mixture of strangeness +1 and −1. The important thing is that these particles will not 'remember' their past history. The fact that they had once passed through an S = +1 filter is immaterial.

By effectively measuring the lifetime of the particles, we have knocked them into an *eigenstate of lifetime* with a 50–50 mixture of strangeness. We now have exact knowledge of the lifetime, but have lost the information about strangeness.

Statistically, half the surviving kaons will emerge from the second S = +1 filter. The strangeness identity will be re-established for these particles and the lifetime status forgotten (they will be in an *eigenstate of strangeness* with a 50–50 mixture of lifetime). There is a 50% chance that they will decay quickly, as a K_s^0. The short-lived kaons will have been regenerated!

Knowing it has S=+1 has changed its life expectancy.

The observation of such regeneration would confirm the existence of oscillations between eigenstates and hence of two-state quantum mechanical systems.

Pais and Piccioni conclude their paper by saying: '*This striking prediction about the behaviour of the K^0 is in some ways similar to the behaviour of polarized light under similar circumstances*'.[3]

Experimental Proof

The theoretical prediction had, of course, to be verified by experiment. This was more easily said than done, since neutral particles cannot be focussed or directed by electric or magnetic fields.

---

[3] *How to verify experimentally a recent theoretical suggestion that the K^0 meson is a particle mixture*, Physical Review 100, 1487 (1955).

Nevertheless, in 1960, Piccioni and his collaborators at the Lawrence Berkeley Laboratory carried out the experiment at an accelerator called the 'Bevatron'. They observed about 200 decays of short-lived neutral kaons regenerated in a lead and steel plate in a bubble chamber. The theoretical prediction had been verified together with a number of other predictions made by the Gell-Mann–Pais scheme.

Oscillations between one identity and another are typical of quantum mechanics and quite foreign to the household world, where things have an independent existence which does not change just because they are observed.

The Lawrence Berkeley laboratory

By changing imaginary cows and horses to real particles we can reformulate Arianda Chernavska's analogy as follows:

Suppose we pass particles through a gate which only lets through $S = +1$ neutral kaons and rejects all others. Next we pass these kaons through a second gate which lets through only long-lived particles and rejects all short-lived ones. Only particles which have $S = +1$ and are long-lived can pass through both gates. To our surprise, approximately half of the surviving particles turn out to be short lived!

Murray Gell-Mann

Murray Gell-Mann was born in New York City on September 15 1929. His parents were immigrants from Czernowitz, in the

Murray Gell-Mann.
Courtesy of NASA

Ukraine. A child prodigy, he tried to teach himself calculus at the age of nine. He also showed an interest in linguistics and archaeology but, by his own account, did poorly in his high school course in physics. Nevertheless, when he won a scholarship to Yale University, he decided to study physics and soon become captivated by the subject.

Gell-Mann may be described as a character who is 'larger than life'. He won the Nobel Prize in 1969 for his work on the theory of elementary particles but he has a very wide spectrum of interests outside physics. In 1984, he co-founded the Santa Fe Research Institute in Interdisciplinary Study which aims to foster a multidisciplinary research community and serve as '*an institute without walls*'. There he leads research into the evolution of human languages with the aim of grouping the known families of human languages into fewer and larger super-families and, ultimately, to trace their origin to a single, hypothetical proto-language.

A glimpse of a wild jaguar while travelling through rain forests in eastern Ecuador prompted him to relate his thoughts on a wide range of topics in the book entitled *The Quark and the Jaguar: Adventures in the Simple and the Complex*.[4]

---

[4] Murray Gell-Mann. *The Quark and Jaguar: Adventures in the Simple and the Complex.* W.H. Freeman & Company (1994).

Chapter 13

Paul Dirac —
Tying Things Together

Paul Adrien Maurice Dirac (1902–1984) obtained a scholarship to study at Bristol University. He chose to follow his brother, Felix, into electrical engineering since he thought that his first love, mathematics, would probably lead to a career in teaching.

As part of his engineering degree, Dirac was assigned to an electrical company as a summer trainee. He received a report which described him as '*a positive menace in the Electrical Test Department ...*' Despite this adverse report, Dirac graduated with a first class honours degree, only to find that he was unable to get a job because of the depression.

Paul Dirac. © Peter Lofts Photography/National Portrait Gallery, London

In 1919, The Times of London reported that British astronomers had detected the bending of light by the sun, just as predicted by Einstein's theory of general relativity and Albert Einstein became a celebrity overnight. The event had a tremendous effect on the young Dirac. He was persuaded to do a mathematics degree in Bristol by the mathematician Ronald Hassé, and graduated just two years later with a first class honours degree. He gained a scholarship to St John's College Cambridge in 1923 to do a PhD in mathematics.

Dirac received a proof copy of Heisenberg's paper from his PhD supervisor. He abandoned his work on Bohr's theory and, in the final year of his postgraduate studies, he developed a new approach to Heisenberg's quantum mechanics. While most physicists were mystified by the mathematics, Dirac thrived on it. He became very excited when he realized that Heisenberg's strange multiplication rule '*provided the key to the whole mystery*'.[1] Heisenberg had used experimental results to guide him to the formulation of his theory, whereas Dirac's approach was more abstract. He worked 'from the top down', beginning with a precise mathematical formulation and only considering experimental predictions afterwards. This was in keeping with his fundamental belief '*that the laws of nature should be expressed in beautiful equations*'.

While he was writing his thesis, Dirac became aware of de Broglie's waves but thought they were just a mathematical curiosity and did not take them seriously. As he was nearing completion of his PhD, he heard that Schrödinger *had* taken them seriously and had developed a *wave mechanics* which appeared to be completely different from Heisenberg's scheme. Dirac did not mention wave mechanics in his thesis, but his interest was stimulated by what appeared to be another 'beautiful equation' to express the laws of nature. As he said later, however, he felt a bit annoyed. '*If we have one good theory, that is all we really want; this was rather too much, an excess of riches*'.

It did not take long for Dirac to acquire a high reputation at Cambridge and, in the autumn of 1926, he received a prestigious scholarship from the 1851 Commission which took him to the centres for quantum theory in Europe. He spent the first six months at Bohr's Institute in Copenhagen, as Heisenberg had done before him. In Copenhagen, Dirac was viewed as extremely eccentric (even by academic standards!). However, in spite of maintaining his customary solitary existence, he seemed to thrive in the easy-going environment of the institute; to say his visit was successful academically would be something of an understatement. He also made new acquaintances, among them were Heisenberg, whom he liked and held in high esteem, and Pauli. Bohr described Dirac's manner during his visit: '*in Copenhagen* [we] *expect anything of an Englishman*'.

---

[1] P.A.M. Dirac. 'Recollections of an Exciting Era' in C. Weiner, ed., *History of Twentieth-Century Physics*. Academic Press, New York. 1977.

Dirac added his own 'slant' to the informal discussions at the institute. His basic attitude was that physical laws can be expressed with true clarity only by mathematics; words and philosophical arguments are inadequate.

During his postgraduate years, Dirac had attended 'mathematical tea parties' held at week-ends by Henry Frederick Baker (1866–1956), who held the unusual title of Professor of Astronomy and Geometry. For Dirac, the tea parties had been the highlight of the week and stimulated his appreciation of the beauty of mathematics for its own sake, with no concern about applications to physical reality. Now, as he thought about the mathematical formalisms of quantum theory, an idea came to him from something he had contributed to the weekly gatherings.

Vector Algebra

At one of the 'parties', he had presented his first ever seminar, which was based on the work of the German mathematician Hermann Grassmann (1809–1877). In 1844, Grassmann had published *Die Lineare Ausdehnungslehre,*[2] a mathematical formalism which pre-dates what is now known as *vector algebra.* Grassmann received little or no recognition at the time, probably because his ideas were too new. Dirac, on the contrary, had not only acquired a deep appreciation of the work for its own sake, but felt that the logic and coherence of vector algebra might very well be reflected in the laws of physics at the most fundamental level.

Hermann Grassmann

Vector algebra is an amalgamation of geometry and algebra, and combines the advantages of both disciplines in a new branch of mathematics. Geometry deals with relations in space while algebra

---

[2] 'The theory of extensive magnitudes'.

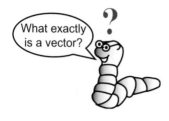

provides methods of logical deduction using equations.

In navigation, the term *vector* is used to indicate a course from place of origin to destination. This expresses the same idea as its mathematical equivalent, a quantity which has both magnitude and direction.

Vectors are usually represented by an arrow, which helps to visualize geometrical operations such as changes in length or direction or the addition of vectors to make a new 'combined' vector. Vectors can not only be added and subtracted but also multiplied by other vectors according to certain rules.

Vectors 'live' in a space which can have any number of *dimensions*. In two dimensions, such as on a flat sheet of paper, we define two independent directions by perpendicular axes which then provide a reference framework. The *projections* of the vector on each axis are called the *vector components*.

Transforming a Vector

We will take, as an example, the vector used to indicate the course from Dublin, Ireland, to Edinburgh, Scotland, a journey of 352 km in a north easterly direction 'as the crow flies'. In the left hand diagram, the components a and b are the *coordinates* of the vector

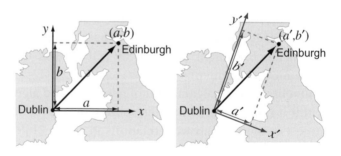

The same vector but different reference frames

(starting from Dublin and ending in Edinburgh) in *vector space*. In this particular example, vector space happens to be real space but this need not be the case.

A characteristic feature of vector algebra is that it is possible to 'look at the same vector from another point of view' by making a *transformation* to another frame of reference.

The right-hand side of the diagram above shows the same physical space. The vector itself is unchanged (Edinburgh is still the same distance and direction from Dublin) but the axes have been rotated, and the projections of the vector on the new axes have the new values of a' and b'. The relationship between the new and the old coordinates depends on the angle of rotation of the axes (about 20° clockwise in this case).

The vector always forms the hypotenuse of a right angled triangle and the projections form the other two sides. As Pythagoras proved in about 500 BC, the sum of the squares of the components remains constant.[3] This is crucial to Dirac's vector representation of quantum mechanics.

In the familiar three dimensions of everyday life, we can still represent the vector by an arrow, this time with *three* components. In the diagram, we visualize a vector which starts from the corner of a room and is illuminated by a light on the ceiling. The shadow of the vector is projected onto the floor and it in turn can be resolved into two components along the edges of the floor. Thus in three dimensions the vector is the resultant of three perpendicular components, two on the floor and the third directed up the corner of the wall.

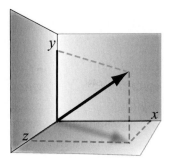

Vector in three-dimensional space

---

[3] It is believed the ancient Babylonians and Indians were aware of the relation even earlier.

Dirac's Representation of a Physical System

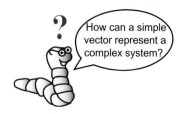

How can a simple vector represent a complex system?

In Dirac's model, the state of a physical system, which can be as simple as a single particle, or as complex as an atom, is represented by a *state vector* which lives in a complex vector space. This space is an abstract mathematical construct and has nothing to do with normal physical space.

The coordinate axes represent the eigenstates of a physical observable of the system, such as position or energy. Dirac's vector space therefore has as many dimensions as there are eigenstates.

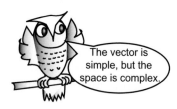

The vector is simple, but the space is complex.

Before the observable is measured, the projections of the state vector on each axis give the *probability amplitude* for that state. The physical meaning is that the square of this amplitude gives the probability that the observable has that value.[4]

After measurement, the system is found to be in some particular eigenstate and the state vector points along the corresponding axis.

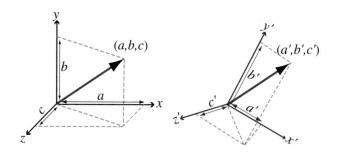

Rotating the axes in three-dimensional space

---

[4] Since the amplitudes form the sides of right angle triangles, the sum of their squares add up to the square of the state vector, which we can choose to be of unit length. As a result, the sum of the probabilities is always equal to one. With 100% certainty, the system must be in one of the eigenstates.

As the model began to take shape, Dirac became very pleased with it. The superposition of quantum states fitted naturally into the scheme. In a sense, the components of any vector exist within the vector, but only become apparent when a frame of reference has been chosen. In a different frame of reference, the vector stays the same but the components along the new axes are different. Similarly the eigenstates of an observable can be said to be 'within' a quantum mechanical system. The probability amplitudes depend on the choice of observable.

Dirac soon realized that the wave mechanics of Schrödinger and Heisenberg's matrix mechanics were special cases of his general theory. In wave mechanics, the square of the wave function gives the probability of finding the particle in a given place. It corresponds to Dirac's vector representation when the eigenfunctions of position form the frame of reference. In the Heisenberg theory, the energy eigenfunctions form the frame of reference, providing the probability of finding that the system has a particular value of energy.

At a lecture given at the 1977 European Conference on Particle Physics in Budapest, Dirac recalled:

> *It was then just a question of a mathematical transformation to pass from the Schrödinger theory to the Heisenberg theory. They were two mathematically equivalent theories for the same underlying physics. That underlying physics is what we now call quantum mechanics.*

Dirac's quantum mechanics provides a general method of finding the possible values of other observables by transforming to the frame of reference made up of eigenfunctions of that observable.

In December 1925, Max Born received the pre-print of a paper entitled *The Fundamental Equations of Quantum Mechanics* and written by Paul Dirac, an 1851 Exhibition Senior Research Student from Cambridge. Born realized that Dirac's theory of quantum mechanics was similar to the Born–Heisenberg–Jordan theory, only just completed. He immediately recognized that, although the author was a '*youngster ... everything was perfect in its way and admirable'.*[5] Dirac had in fact submitted his work nine days before Born and his collaborators. It was a remarkable achievement for the young man, who had worked alone and without the sort of support that was taken for granted at centres such as Göttingen.

Dirac arrived in Göttingen in early 1927, having just submitted the paper *Electrons and Fields* to the Royal Society of London. The atmosphere in Göttingen was very different from Bohr's institute. Many of the professors maintained an exalted position but Max Born was different and Dirac benefitted from the environment he created. Dirac remained polite but aloof and many German scientists were alienated by his manner, to the extent of applying the words 'difficult' and 'overrated' to Dirac and his work.

Dirac's manner was often misunderstood, and not only in Germany. He was economical with his words, and not everyone realized he did not mean to be rude. One can understand how he might have given the wrong impression from this anecdote relating to a lecture given in Madison, Wisconsin.

Question from the audience: '*I don't understand the equation on the top-right hand of the blackboard.*'

Dirac remained silent, looking unconcerned, until the chairman asked him to reply.

Reply: '*That was not a question, it was a statement.*'[6]

Uniting Quantum Mechanics with Einstein's Theory of Relativity

Schrödinger's theory was not consistent with Einstein's *special theory of relativity* mainly because Schrödinger used Newton's classical formula

[5] Nancy Greenspan. *The End of the Certain World. The Life and Science of Max Born: The Nobel Physicist Who Ignited the Quantum Revolution.* John Wiley, Chichester. 2005.
[6] Leopold Infeld. *Quest: The evolution of a scientist.* The Scientific Book Club, London. 1941.

for energy. This formula does not take into account mass energy, which is an essential ingredient of the atomic and nuclear world.

Oskar Klein (1894–1977) and Walter Gordon (1893–1939) proposed a solution to this problem which was generally considered satisfactory for Schrödinger's wave mechanics. Dirac, however, did not agree:

> *Most physicists were happy with this development of the Klein–Gordon equation. They said, here you have a good relativistic quantum theory. But I was most unhappy ...*

The reason for Dirac's discontent was that one could not apply transformation theory to the new equation. He worried about this problem until the end of 1927, and then the solution came to him almost by accident. The problem was technical rather than physical. By using matrices rather than simple algebraic variables (*Dirac 4 × 4 matrices*), he could make the relativistic expression for the energy of the electron compatible with his transformation theory. In a classic paper *The Quantum Theory of the Electron,* published in the Proceedings of the Royal Society of London in 1928, Dirac derived his wave equation for the electron; it is compatible with both Einstein's theory of relativity and his own transformation theory.

An Equation with Something to Tell

There was just one difficulty with the equation and it appeared to be minor; the equation gave two solutions for the kinetic energy of an electron, and one of them was negative. Negative kinetic energy does not make sense. In Dirac's words:

> *It corresponds to electrons with a very peculiar motion such that the*

Chess board

faster they move, the less energy they have, and one must put energy into them to bring them to rest.

This sort of situation is not unusual in mathematics and it is normal to dismiss such negative solutions as 'non-physical'. Given that a chess board is divided into 64 equal squares, one can deduce that the number of squares in each row is the square root of 64, which is 8. If we want to be pedantic, the answer is ±8, but the negative solution does not make sense and is discarded.

Dirac was not ready to dismiss the negative solutions of his wave equation for the electron 'out of hand'. It was as if the equation was cleverer than he, and was trying to tell him something. Perhaps the negative sign was a code for another particle? Dirac did not dare to postulate the existence of a new particle. The climate of opinion at the time did not favour new particles. There were just two particles, the electron and the proton.[7] It was believed that atoms and therefore all matter were made of these two fundamental building blocks. Evidence much more convincing than the negative solution of a quadratic equation was required to perturb such a simple and complete scheme.

For a while Dirac thought that the other particle might be the proton, but it soon became clear that the proton, with a mass about 2,000 times greater than the electron, could not be 'twinned' with the electron. Mathematically, what was needed was a 'mirror image' particle with the same mass as the electron but positively charged. Just as it seemed that the theory had come to a dead end, exciting news came from America.

A New Particle from the New World

On 15 March 1933 Carl Anderson (1905–1991), published a paper in the American journal *Physical Review* reporting 1,300 photographs of cosmic-ray tracks taken in a Wilson cloud chamber at the California Institute of Technology. An external magnetic field caused curvature

---

[7] The neutron was not discovered until 1932.

of the tracks which helped to identify the particles. Among these
photographs, Anderson found 15
tracks of positive particles which
were too curved to have a mass as
large as the proton. The paper was
illustrated with the photograph
reproduced below.

What appears, at first, to be an
innocuous single track passing
through a 6 mm lead plate close
to the centre of the chamber. We
can deduce that the particle was
travelling downwards since the

*Track due to a positron
in a cloud chamber*

curvature is greater after the particle has been slowed down in the
lead plate. The direction of the curvature shows that the charge is
positive. The author concludes:

> *From an examination of the energy-loss and ionization produced it is con-
> cluded that the charge is less than twice, and is probably exactly equal to,
> that of the proton. If these particles carry unit positive charge the curvatures
> and ionizations produced require the mass to be less than twenty times the
> electron mass.*

The result was published in the *Science Newsletter* of 19 December
1931. The editor suggested the name *positron* for the new particle;
Anderson did not much like it, but agreed nevertheless.[8]

A few months later, the existence of the positron was confirmed by Blackett and Giuseppe Occhialini (1907–1993) at the
Cavendish Laboratory in Cambridge. They used their newly developed trigger system to expand the cloud chamber when a cosmic
ray passed through Geiger counters above and below the chamber.
This increased the efficiency of the experiment many times over,
and they obtained many photographs showing tracks due to both

[8] Abraham Pais. *Inward Bound.* Oxford University Press, Oxford. 1986.

positively and negatively charged particles with the mass of an electron. Blackett, always a cautious man, discussed his experimental results with Dirac. Reassured, he went ahead with the *positron* interpretation.

When Blackett announced his result, there was great excitement and wide press coverage, showing pictures of pair production. Dirac was not at the presentation. Characteristically, he chose to avoid the limelight and did not put in an appearance; he was giving a seminar in another part of the building.

A prediction made from pure mathematics was proven. It came from the interweaving of two theories; relativity, born in the laboratory of the mind of Einstein, and quantum theory, born in the minds of those who worked in the 'quantum laboratory', where Dirac himself had played no small part. Anderson's discovery of the positron was fortuitous; he was not 'looking for' Dirac's mirror image particle. As he said subsequently: '... *But I was not familiar in detail with Dirac's work. I was too busy operating this piece of equipment to have much time to read his papers.*'

Antimatter

The electron is not the only particle with a twin, a special kind of mirror image replica of itself. The image is opposite in every parameter: opposite charge, opposite spin, even opposite mass, in the sense that if the twins meet they mutually annihilate. This kind of matter has been given the name *antimatter*.

For every particle there is an antiparticle. The *antiproton* was discovered in 1955 by Emilio Segrè (1905–1989) and Owen Chamberlain (1920–2006) and their team at the Lawrence Radiation Laboratory in Berkeley. One year later came the discovery of the *antineutron* by Bruce Cork (1916–1994) and his team, also at Berkeley.

In 1965, the first 'antimatter nuclei' were produced. They were *antideuterons*, each consisting of an antiproton and antineutron stuck together. They were created simultaneously by two teams of physicists, one led by Antonino Zichichi (1929–), using the Proton Synchrotron at the European Organization for Nuclear Research (CERN), and the other led by Leon Lederman (1922–), using the Alternating Gradient Synchrotron accelerator at the Brookhaven National Laboratory, New York.

Antiprotons Made to Order

Antimatter was first observed in cosmic radiation, and later from decays of radioactive isotopes. The next stage was to produce beams of antiparticles at accelerators. In 1982, an antiproton storage ring was built at CERN to accumulate antiprotons and then store them for injection into the Super Proton Synchrotron (SPS) accelerator, where at least some of them would collide head-on with protons.

Atoms of Antihydrogen

The next step in the study of antimatter was to create a structure resembling a 'regular' atom, but with the nucleus and the electrons replaced by their corresponding antiparticles. Atoms of antihydrogen were observed at CERN in 1995 and then at Fermilab, near Chicago, in 1997. In both instances, they were produced in flight, moving at almost the speed of light which seriously limited the experiments which could be done with them.

Then, in August 2002, CERN announced the first successful production of atoms of *antihydrogen* (positrons and antiprotons joined up as atoms) in a 'regular' antimatter environment. The task here was to slow them down rather than accelerate them. Antiprotons from the storage ring were fed into another ring where they were decelerated and then trapped in an 'electromagnetic cage'. They were then mixed with positrons from the radioactive decay of sodium 22, and held in a second trap. They were now moving slowly enough to have a significant chance of joining up as atoms. In the first few weeks of the experiment the team estimated that about 50,000 antihydrogen atoms had been produced. (This may sound like a big number, but it would have to be multiplied by 10^{18} — a billion billion — to make 1 g of antihydrogen.)

Atoms of antimatter are exceptionally difficult to store; the vast majority of them meet 'ordinary' matter and mutually annihilate within a tiny fraction of a second. In May 2011, CERN announced that 309 atoms of antihydrogen had been held in a trap for 17 minutes. This opens up the possibility of creating a small lump of antimatter and allowing it to fall (or rise) under gravity. Such an experiment would settle the question of whether antimatter is attracted or repelled by ordinary matter.

Even more interesting is the prospect of studying the spectrum of antihydrogen, to determine whether or not it is identical to the spectrum of hydrogen. Apart from being of great theoretical interest for its own sake, it could answer the speculation with which Dirac concluded his Nobel lecture in 1933:

> *It is quite possible that for some of the stars it is the other way about, these stars being built up mainly of positrons and negative protons. In fact, there may be half the stars of each kind. The two kinds of stars would both show exactly the same spectra, and there would be no way of distinguishing them by present astronomical methods.*

Dirac, the youngest theoretician to have been awarded a Nobel Prize, wrote of Niels Bohr:

> *I feel that all my deepest ideas have been very greatly and favourably influenced by the talks I have had with you, more than with anyone else.*[9]

Symmetry

Throughout the ages, *symmetry* has been associated with beauty and perfection. Symmetry is inherent in painting and architecture both in ancient and in modern times. The mausoleum of the Taj Mahal

---

[9] Letter from Dirac to Bohr. 28 November 1933.

in Agra, India, is a perfect example; it was built by Mughal
emperor Shah Jahan in memory
of his favourite wife, Mumtaz
Mahal, and is considered as one of
the new Seven Wonders of the
World. Its beauty lies in a symmetry which however is not absolute.
A perfect symmetry would be a
sphere which looks the same from
any direction. The artistry of the
Taj Mahal lies in the blending of
symmetry with diversity.

*Taj Mahal. © Courtesy of Sandeep
Dhirad, Wikimedia Commons*

Richard Feynman, in his book *Lectures on Physics,*[10] tells the story
of a gate in Neiko, Japan, which is known as the most beautiful gate
in all Japan. The two sides of the gate are covered with beautiful
carvings in perfect symmetry except for one small dragon's head,
which is upside down. The story goes that this was done deliberately
so that the gods would not become jealous of man's perfection.

Paradoxically, that imperfect dragon adds interest to the famous
gate. Excessive symmetry is monotonous; a complete vacuum has
perfect symmetry, but is devoid of any features. Something must
appear, something must happen to break the symmetry and make it
interesting.

The Birth of the Universe

In the early universe, fractions of a second after the *Big Bang*, particles and antiparticles were being created and then annihilated back
into energy continuously. After a few seconds the temperature
dropped; there was not enough energy to create matter, but particles and antiparticles continued to annihilate one another. Their
mass energy was converted into electromagnetic radiation, which is

---

[10] Richard P. Feynman, Robert B. Leighton and Matthew Sands. *The Feynman Lectures
On Physics.* Addison Wesley Publishing Company, New York. 1985.

still with us today as the *cosmic microwave background radiation* discovered by Arno Penzias and Robert Wilson in 1965.

At the end of this battle of attrition, all the antimatter was annihilated but some matter was left over. We do not know the reason for this imbalance, but we do know that without it we would not be here. We are made out of this surplus matter and so, as far as we know, is the rest of the visible universe.

Symmetry in Fundamental Interactions

Human society has introduced conventions which make a distinction between left and right. Most cultures use the right hand to greet one another, some drive on the left, some on the right, corkscrews and nuts and bolts usually have right-hand threads. The hands of a clock turn in a clockwise direction. All these rules are conventional; it would be equally possible to reverse left and right or make a mirror image of a clock which would go anticlockwise. Such a clock, apart from being unconventional would work in exactly the same way and 'keep time' just as well as the 'normal' one.

For centuries it was taken for granted that in nature there are no such conventions and that there is complete symmetry between left and right. It seemed obvious that the same experiment carried out on an object and its mirror image should yield the same result.

In May 1956, Chinese-American theoretical physicists, Tsung-Dao Lee from Columbia University New York and Chen Ning Yang from the Institute for Advanced Study in Princeton decided to carefully review all known experiments involving so-called *weak nuclear interactions*. After several weeks they came to the conclusion that none of them had any bearing on the question of symmetry under mirror reflection (or, in the language of quantum mechanics, *the conservation of parity*).

Tsung Dao Lee (1926–) and Chen Ning Yang (1922–) immediately submitted a paper[11] in which they suggested possible experiments to test the conservation of parity. The response from experimentalists was not enthusiastic; most of them thought it was a waste of time and resources. The exception was Madame Chien-Shiung Wu, a professor at the University of Columbia, who was a good friend of Lee and Yang.

Radioactive Decays Distinguish Right from Left

Wu's experiment involved nuclei of cobalt 60, which decay radioactively by emitting high-speed electrons. Each nucleus has an intrinsic spin, which means it acts like a tiny magnet. By placing a sample of Co^{60} in a strong magnetic field, the nuclei are oriented so that their spins are in the same direction, say anticlockwise, when viewed from above. If left-right symmetry is obeyed there should be no significant difference between the numbers of electrons emitted in the upward and downward directions.

An experiment confirms parity violation

To minimize the level of thermal agitation, the system had to be maintained at a very low temperature, so the experiment was conducted at the Cryogenics Physics Laboratory at the National Bureau of Standards in Washington. Parity violation was first observed on 27 December 1957; the results are summarized in the schematic diagram.

---

[11] T.D. Lee and C.N. Yang. *Question of Parity Conservation in Weak Interactions.* Physical Review 104, 254–258 (1956).

In the 'real' world, when cobalt 60 nuclei are spinning anti-clockwise, as seen from above, more electrons are emitted in the upward than in the downward direction. The situation depicted in an imaginary 'mirror world' does not exist. Lee and Yang were right in their suspicion. There is a significant asymmetry. Nature had sprung the biggest surprise since Planck's discovery of the quantum; she had distinguished between left and right. (It may not have been such a surprise to Dirac, who had written of the possibility as early as 1949.)

As Yang recalled at the Nobel presentation in December 1957:

> *the fact that parity conservation in the weak interactions was believed for so long without experimental support was very startling. But what was more startling was the prospect that a space-time symmetry law which the physicists have learned so well may be violated. This prospect did not appeal to us.*

Physicists had to adjust their thinking and look for a deeper symmetry. For a number of years they supported the theory, first proposed by the Russian physicist Lev Landau in 1957, that the true symmetry of nature is charge/parity (CP) symmetry which states that the laws of physics should be the same if a particle were interchanged with its antiparticle (charge conjugation), and left and right were swapped (parity exchange).

In 1964, a study of the decays of neutral kaons (the particles that we met in the previous chapter) by James Cronin (born in 1931) and Val Fitch (born in 1923) showed that CP symmetry is broken, but only at the level of three decays in 1,000. This tiny amount of asymmetry could explain the dominance of matter over antimatter in the universe.

Dirac summed up the philosophy of quantum mechanics when he addressed delegates at the Indian Science Congress in Baroda in 1955:

> *When you ask what are electrons and protons I ought to answer that this question is not a profitable one to ask and does not have a meaning. The*

important thing about electrons and protons is not what they are but how they behave — how they move. I can describe the situation by comparing it to the game of chess. In chess, we have various chessmen, kings, knights, pawns and so on. If you ask what a chessman is, the answer would be [that] it is a piece of wood, or a piece of ivory, or perhaps just a sign written on paper, [or anything whatever]. It does not matter. Each chessman has a characteristic way of moving and this is what matters about it. The whole game of chess follows from this way of moving the chessmen.

He had found the knack of communicating his science to the public.

In 1969, Dirac retired from Cambridge, one of the world's most highly ranked physics departments. He had been the Lucasian Professor of Physics since he was 29 years old. In January 1971, he accepted the position of Visiting Emeritus Professor at a university whose physics department was ranked number 83 in the USA. When the move to appoint Dirac was first suggested, there was some unease at the idea of recruiting such an old man (Dirac was 67 at the time). Opposition was effectively silenced by the head of department when he announced to the faculty:

To have Dirac here would be like the English faculty recruiting Shakespeare.[12]

[12] Graham Farmelo. *The Strangest Man. The Hidden Life of Paul Dirac, Quantum Genius*. Faber and Faber, London. 2009.

Chapter 14

Richard Feynman — The Strange Theory of Light and Matter

Richard Feynman (1918–1988) was a character like no other. A brilliant theoretical physicist, he had the ability to express the most difficult concepts in words that were simple and conveyed his unique insight into the beauty of the laws of nature. Hearing him speak left one inspired and entertained, with the erroneous impression that one understood the deepest mysteries of physics. All this from someone who, as a child, had been slow in learning to speak; by the age of three he had not spoken a single word.

Feynman was born on 11 May in Far Rockaway, a suburb of New York City in the borough of Queens. As an undergraduate at the Massachusetts Institute of Technology (MIT), he studied the standard course and, at the same time, attended research seminars where he soon realized that the fundamental problem of the day was associated with the quantum theory of electricity and magnetism. Dirac's first book, published in 1930,[1] concludes with the remark that the point-charge model of an electric charge already involves some difficulties in classical theory, and that it is therefore not surprising that the passage into quantum theory should bring further difficulties. The task of resolving these difficulties presented Feynman with a challenge and an inspiration.

---

[1] P.A.M. Dirac. *The Principles of Quantum Mechanics*. Clarendon Press, Oxford, 1930.

The main problem lies in something which does not immediately spring to mind — the interaction of the electron with its own electric field. This field, just like the field of any 'point' charge, diverges and decreases inversely as the square of the distance from the charge. Very close to the charge, the field is extremely high. For a true point charge, such as the electron, the field increases without limit as it converges back towards the charge. This leads to an infinite energy of interaction of the electron with itself.

There were many questions to be answered. Perhaps electrons cannot act on themselves and can act only on other electrons? This makes little sense when dealing with a field created by many electrons. Each one contributes to making that common field; does this field then act differently on every electron? How could such an idea be reconciled with well-established laws of electricity and magnetism?

In 1939, Feynman entered graduate school at the University of Princeton, where his research supervisor was a young professor, John Archibald Wheeler (1911–2008). Under Wheeler's guidance, Feynman spent many months trying various approaches to a quantum theory of the interaction between light and electric charges, which became known as quantum electrodynamics (QED). Dirac's words echoed in Feynman's mind: '*It seems that some essentially new physical ideas are here needed*'.[2]

Feynman's Thesis

In 1942 Feynman submitted his PhD thesis: *The Principle of Least Action in Quantum Mechanics*. Feynman had first heard of the principle of least action from Abram Baden, his teacher in Far Rockaway. It seemed logical to Feynman that, if such a principle forms the basis of classical mechanics, it must also apply in the quantum world. His thesis set out to approach quantum mechanics from a different perspective, using the methods introduced by Lagrange and his peers about 200 years earlier. In this way, the problems encountered when considering a physical process step by step might be avoided by starting from a global statement about the entire process.

---

[2] Richard P. Feynman. Nobel lecture. 1965.

The 'Crazy Ideas' of Quantum Mechanics

Leonard Mautner and Richard Feynman had been good friends since their school days in Far Rockaway. Mautner's wife Alix, whom Feynman also had known since childhood, took up an academic career and later became Professor of English at the California State University, in Los Angeles. She had often asked Feynman to explain 'the crazy ideas of quantum mechanics' and, for many years, had tried to persuade him to give a series of talks on 'the physics of small particles' that would be understandable to her and to others outside the physics community. Sadly, Alix died in 1982 and, perhaps feeling guilty that he had not done so earlier, Feynman prepared the first series of *Alix G. Mautner Memorial Lectures* in her honour.

Feynman gave these lectures in New Zealand which, as he said, was far enough away in case they did not work out! On the contrary, the lectures were very successful and were subsequently published in a book entitled: *QED, The strange theory of light and matter.*[3]

A 'Mind-Boggling' Problem with Shop Windows

In keeping with the intended style, Feynman's book does not begin with the complex mathematical problem of infinite energies, but rather as follows:

> While partial reflection (of light) by a single surface is a deep mystery and a difficult problem, partial reflection by two or more surfaces is absolutely mind-boggling.

Feynman is referring to something familiar; we might call it 'the shop-window' effect. When light falls on a glass surface most of it is transmitted and enters the glass, but a small fraction is reflected. This is why a shop window, made for the purpose of displaying the goods inside, can also be used as a mirror.

---

[3] Richard P. Feynman. *QED, the strange theory of light and matter.* Princeton University Press, New Jersey. 1985.

The Partial Reflection of Photons

As soon as we start thinking in terms of photons, new questions and new inferences arise. We will begin with our commonly experienced shop window scenario and examine the effect methodically, using two idealized experiments based on our knowledge of the household world.

Reflection at a Single Surface

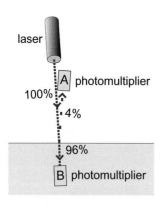

Reflection at a single surface

Starting systematically, we send a thin pencil beam of light from a laser onto a glass surface. The beam is very dim so that the photons, which are identical, arrive one by one. Most of them enter the glass, the remaining few are reflected. A photomultiplier is placed at a point A to count the reflected photons, while a second photomultiplier at B, inside the glass, counts the photons which enter the glass, as illustrated in the figure. (This is an idealized experiment, which means that we will not worry about how to embed a photomultiplier in glass.)

From classical experiments on light, we expect that about 4% of the photons will be reflected, while the remaining 96% will be transmitted. Although this confirms our experience at shop windows, it does require an explanation.

All photons are identical; they enter the surface in the same place and under exactly the same circumstances. What Feynman found mysterious was that some photons are reflected, but not others. A photon cannot split into two parts, it must either continue or turn back. How does the photon 'know' which option to take?

We can construct various theories to rationalize partial reflection. A reasonable theory might be that the phenomenon has something to do with the glass. Perhaps the atoms at the surface form a fine mesh of atomic dimensions, which acts as some sort of combined filter and reflector? If so, we still have to explain (among other things) why this only happens at the surface and why the photons have a free passage once they are inside the glass.

Reflection at Two Surfaces

Another way to test this theory is to do a second experiment, in which we replace the solid glass block with a sheet of glass and put the second photomultiplier below it. The photons which are transmitted at the upper surface continue on through the glass and eventually arrive at the lower surface as shown below. Some will be reflected back into the glass, the rest will continue onwards. The photomultiplier at A will now count the photons reflected at both surfaces.

What Our 'Theory' Predicts

According to our theory, every boundary between glass and air acts as a 'filter'; we expect 4% reflection and 96% transmission at both upper and lower surfaces, no matter whether the photons are going from air to glass or from glass to air. This applies to the photons arriving at the lower surface and then again to any which return to the upper surface. We can predict that out of every 100 photons in the original beam marginally fewer than eight should return to the photomultiplier at A. This is just a quick estimate, in which we have ignored the decrease in the intensity of the beam and also further reflections along the glass. It is not necessary to

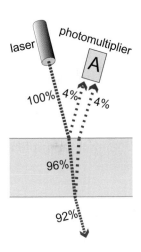

Reflection at two surfaces

make an exact calculation, because this prediction bears no relation to the experimental result. The theory is simply wrong.

Photons with Different Properties?

We could propose an alternate theory. Perhaps it is something to do with the photons themselves? Maybe there are different kinds of photons, those which cross the surface and those which are reflected. According to this, the photons which penetrate the glass and arrive at the lower surface should also cross that surface and escape back into the air. The second surface makes no difference; no additional photons will return to A and the number counted remains at 4%. This theory is also wrong.

The Unlikely Result of a Real-Life Experiment

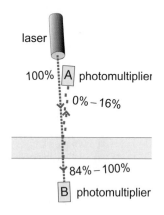

Experimental results

Nothing could prepare us for the results of a real-life experiment. When we actually count the number of photons reflected from different sheets of glass, we get apparently random outcomes. Sometimes no photons are reflected and sometimes as many as 16 out of 100 return to photo-multiplier A. The experience turns out to be particularly frustrating. It is impossible to reproduce consistent results.

We could explain what is happening in terms of a wave model, but here we are definitely dealing with individual particles. The photomultipliers give an audible click as the photons arrive one by one.

The Answer to the Question of Partial Reflection

In our experiments, the source is so faint that if the photons were to travel in a straight line, they would be separated by thousands of miles. In this case when any particular photon arrives at the glass, the next photon has not even been produced by the laser. Each photon is on its own and must 'decide' whether to enter the glass or turn back.

If we insert a third plate of glass, or any number of additional plates, the number of reflected photons will change. What Feynman found mind-boggling was that what lies ahead in space and time can have an effect on the fate of the photon when it arrives at the glass.

The amount of partial reflection is not related simply to what happens at the first surface, but depends on 'the road ahead'.

In Feynman's words:

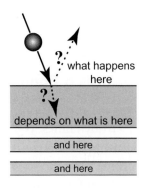

How does the photon know?

> *We find ourselves chasing down through surface after surface, wondering if we have finally reached the last surface. Does the photon have to do that in order finally to 'decide' whether to reflect off the first surface?*

Personifying the photon further, we might say that it looks into the future, explores every possible path and then decides on the most likely action.

In the household world we look into the future but we don't have to explore it ourselves; we can rely on others to advise us about the road ahead, but who advises the photon on what is ahead?

A Chance Meeting

Shortly before submitting his thesis, Feynman chanced to meet Herbert Jehle (1907–1983), a German physicist who had escaped to America in 1941 from a Nazi-controlled internment camp in France. When Jehle heard the subject of Feynman's thesis, he drew his attention to a paper entitled *The Lagrangian in Quantum Mechanics*, which Dirac had published in a little-known journal called *Physikalische Zeitschrift der Sowjetunion* in 1933.[4] The next day they both went to the library to search for the paper. It was there; Dirac had worked on the same idea as Feynman. It was only a start,

---

[4] P.A.M. Dirac. *The Lagrangian in Quantum Mechanics*. Physikalische Zeitschrift der Sowjetunion 3, 64–72 (1933).

but it gave Feynman great encouragement to set out on the road towards a complete theory.

Towards a Complete Theory

This was the beginning of the quest to develop the theory of quantum electrodynamics (QED) for which Feynman, together with Julian Schwinger (1918–1994) and Sin-Itiro Tomonaga (1906–1979), received the Nobel Prize in 1965. All three reached the same final result, but Schwinger and Tomonaga's theory is strictly mathematical whereas Feynman's theory is more physical and can be visualized. Nevertheless his 'sum over histories' approach is still very much contrary to 'common sense' and once prompted him to remark: '*if I could explain it to the average person, I would not have been worth the Nobel Prize*'.[5]

Sum Over Histories

Feynman's grand principle:

> *Anything which might have happened influences that which does happen.*

This means if we want to calculate something like the percentage of photons which will be reflected at a surface, we can't just work on the basis that the photon will take the shortest path; we have to take into consideration any other paths the photon could take. Fortunately for us, this seemingly horrendous task is not quite as daunting as it seems at a first glance. We will start with a simple scenario and build up to the puzzle of partial reflection.

A Photon Goes From Place to Place

The simplest action is that of a photon going from A to B in free space. Each path it could take represents a possible *history*. Feynman lets us imagine a stopwatch, which times the photon as it goes from A to B by every possible path. The stopwatch has a single hand,

---

[5] Commencement address at Caltech. 1974.

which rotates very rapidly. This hand, represented by an arrow, is the *probability amplitude vector.*

In line with his commitment to Alix to make the physics understandable to those outside the physics community, Feynman uses a stop watch, not only to indicate the time difference between various paths, but also the interference of the wave functions of photons and electrons. The mathematics of de Broglie waves and Schrödinger wave functions is contained in the rotating vector of the hand of the stopwatch.

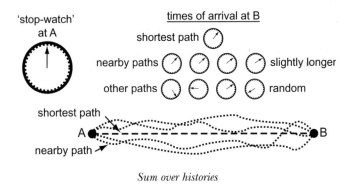

Sum over histories

We start the watch when the photon leaves A and stop it at the instant the photon reaches B, by which time the hand will have made billions of revolutions. We do this for every possible history. The position of the hand at the moment of arrival represents the amplitude vector *for that history.*

The Probability that the 'Event' Will Happen

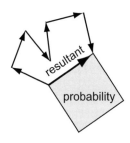

It's the square that counts

When the amplitudes for all paths have been added, the resultant vector representing the *sum of histories* is the *probability amplitude* that the photon from the source at A will arrive at the detector at B. The square of this amplitude gives the *probability* i.e. the fraction of photons which will arrive at the detector. This probability is a quantity we can measure experimentally. Paths very close to the straight line (path of least time) take only slightly longer,[6] and their amplitude vectors lie in almost the same direction, as indicated by the hand of the stopwatch. Only these paths make a significant contribution to the final resultant.

Are we making things too complicated? Instead of making the simple statement '*light goes in a straight line from A to B*', we are now saying 'light can go along an infinite number of paths from A to B, provided they take almost identical times'.

Going to Infinity

How do we add an infinite number of vectors? As we go to more roundabout paths, the amplitude vectors become smaller and smaller and soon we can ignore them. The trick is to know which ones we must count. The sum of these vectors gives us the final amplitude and the probability that the photon will go from the source at A to the detector at B.

---

[6] This is true specifically for the shortest path because a minimum occurs at a *stationary value* of a mathematical function.

Are We Trying to do Easy Things the Hard Way?

At this point it seems relevant to ask what we have achieved with all this business; are we being over-elaborate? Is there a concrete way to prove that these histories are relevant to what actually happens? Let us see what happens if we block paths close to the straight line and thereby eliminate a whole set of possible histories.

The result is quite remarkable. As we gradually make the slit narrower, the light which gets through not only becomes fainter but begins to spread out. We have eliminated a number of possible histories and now the photomultiplier at C some distance

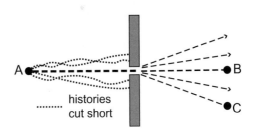

Single slit diffraction explained by missing histories

from the beam direction begins to click. The effect is known as *single slit diffraction*. We have our answer: the possible histories are more than relevant, they are essential.

The 'Mind-Boggling' Experiment

Sometimes the problem can be simplified to just a few paths. A case in point is the reflection of light by two surfaces, the experiment which Feynman described as 'mind-boggling'.

The paths from the laser to the photomultiplier can be broken up into a series of steps. At the beginning, the probability amplitude is of unit length.

At the points where the photon meets the upper and lower surfaces, the length of the vector changes abruptly and the path divides into two branches. Each time, the amplitude for transmission shrinks by a factor of 0.98 and the reflected amplitude is reduced dramatically by a factor of 0.2. When the photon meets the glass for the first time, the probability for transmission is $0.98^2 = 0.96$ or 96% and the probability for reflection is $0.2^2 = 0.04$ or 4%.

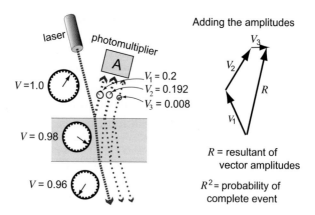

The stopwatches help us to understand

The process continues with multiple reflections and transmissions; each time the path branches out into two, the amplitude for each branch shrinks by the same factor. The size of the vector is constant along the straight line segments; the reduction in amplitude at each surface is shown by the shrinking face of the watch.

Soon, the amplitudes for further paths become negligibly small and can be ignored. We have the critical information, the magnitude and direction of the amplitude vector — the hand of each stopwatch — at the moment of arrival at the photomultiplier.

Feynman is Right!

We are now ready to add the histories. The amplitude vectors of photons arriving at A are V_1, V_2, V_3, etc. In this case, only the first three need to be considered, the others are just too small. These amplitude vectors generally point in different directions and the resultant of the vectors is the probability amplitude for the whole event. It is at this point that the 'mind-boggling' conclusion becomes apparent.

As illustrated in the diagram below, the resultant can be anything from 0 to a maximum value of 0.4 units (when all three happen to be in phase). The corresponding probability ranges from 0 to 16% (the direction of the resultant does not affect the probability).

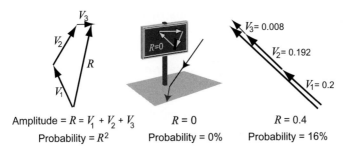

Amplitude = $R = V_1 + V_2 + V_3$ $R = 0$ $R = 0.4$
Probability = R^2 Probability = 0%   Probability = 16%

Feynman shows us how it works!

The relative directions of the amplitude vectors depend on the additional turns made by the stopwatch for each path, which in turn depends on the thickness of the glass. The paths for transmitted rays follow a similar pattern and it turns out that when the probability for reflection is zero, the probability for transmission is 100% and when there is a maximum 16% reflection, transmission is minimum, at 84%, with equivalent matched pairs of values in between. The rules of nature ensure that the accounts balance at all times. The probabilities that the photon will take one or other of these options add up to 100%.

Such is the curious picture presented by Feynman's theory. A single photon arrives at a glass surface. It throws imaginary dice to decide whether to enter the glass or be reflected. Somehow it is informed of the histories that lie ahead and can set the odds appropriately. All the things which could happen if it goes one way or the other play a role in what actually does happen.

Electrons and Photons

What has been presented so far is a glimpse into the quantum electrodynamics of Feynman's Alix G. Mautner Memorial lectures. His purpose was '*to achieve maximum clarity and simplicity without compromise by distortion of the truth*'. Behind this picture lies the deep theory of the interaction of light with matter or, more precisely, of photons with electrons. Electrons fit into the theory according to the same rules of drawing and combining amplitudes. They combine to produce all basic phenomena of nature with the exception of nuclear interactions and gravitation.

> According to QED there are just three basic actions:
>
> A photon goes from place to place.
> An electron goes from place to place.
> An electron emits or absorbs a photon.
>
> These laws describe how atoms interact with each other
> and govern the chemistry of the whole world.

Feynman Diagrams

These three apparently simple actions combine into complex processes which have to be calculated. To quote Feynman: '*It takes seven years — four undergraduate and three graduate — to train our physics students to do that in a tricky, efficient way.*' To set up a framework for such calculations, Feynman invented a form of diagram which bears his name and illustrates the processes in a way that is deceptively simple.

These diagrams bear a certain resemblance to the illustrations that have gone before, but there is an essential difference. Feynman diagrams combine space and time and describe events and paths as points and lines in *space-time*. Each point denotes a certain place and a certain time.

These are the Feynman diagrams for the three basic actions:

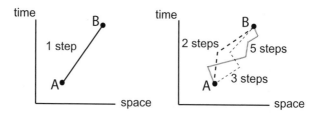

An electron goes from place to place

An electron can go from A to B by the direct path but it can also go along a number of segments, 'hopping' from one intermediate point to another. There is an amplitude vector for each path, and an amplitude for each change in direction. There are an infinite number of paths and each path can change direction at an infinite number of points. All possibilities have to be added to get the resultant amplitude for the electron to go from A to B.

The next diagram illustrates the action in which an electron emits or absorbs a photon. There is a certain amplitude for this to happen which is a measure of the interaction of electrons with photons, the central issue in electrodynamics. This *coupling amplitude* has a fixed numerical value for electrons. Expressed in a more familiar way, it is called the *charge* of an electron.

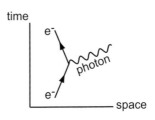

An electron emits or absorbs a photon

The exchange of a photon between two electrons is the process which gives rise to the electromagnetic force between them.

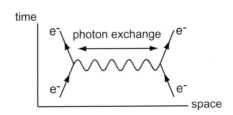

Two electrons exchange a photon

Putting the Three Actions Together

Combinations of these basic actions are represented by Feynman diagrams of varying complexity. They may involve large numbers of electrons and photons and numerous couplings. More than one photon can be exchanged, an electron can emit a photon and later reabsorb the same photon, a photon can disintegrate into an electron-positron pair and so on; the possibilities are unlimited.

The Most Accurate Prediction Ever

According to the theory proposed by Dirac in 1928, the electron possesses an *intrinsic spin*. As a consequence it acts as a tiny magnet whose strength is described in terms of its *magnetic moment.*

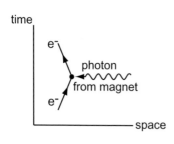

Dirac calculation

Dirac calculated that the electron is exactly twice as effective in creating a magnetic effect as it would be if it were a classical object. His calculation is represented by this diagram on the left which involves one coupling of an electron with a photon.

The diagram below involves three couplings and represents a mechanism first suggested by Schwinger in 1948. An electron emits a photon and later reabsorbs the same photon. In the meantime it absorbs a photon from the magnet. This is less likely than the direct single coupling process, nevertheless it has to be considered and included as a correction factor to the magnetic moment. The correction turns out to be a factor of about 1.00116. In order to calculate it one

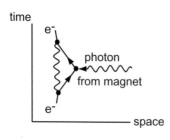

Schwinger calculation

must sum the amplitudes for every point from which the photon can be emitted and every point where it can be absorbed; nevertheless Feynman considered it 'relatively simple', something his students learned to do in the second year of their postgraduate course.

For added accuracy, we need to consider more and more couplings. These may involve the emission and absorption of more photons and exotic processes in which electron-positron pairs are created and annihilated. The computation becomes long, tedious, and complicated, even by Feynman's standards.

With the help of computers, possibilities with seven couplings have been calculated; these involve over 10,000 diagrams with

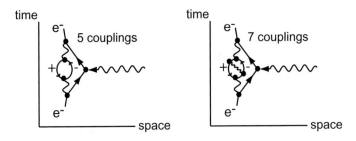

More couplings — greater accuracy

about 500 terms in each diagram. Fortunately there is an end in sight; as the number of couplings increases their contributions become negligible and the final answer converges to a single value.

Experimental Confirmation

There is little point in making highly precise theoretical calculations unless the result can be compared with equally accurate experimental data. Fortunately, the magnetic moment of the electron is one of the most accurately measured physical constants allowing the following comparison:

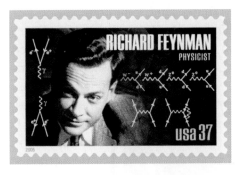

Richard Feynman. Courtesy of the United States Postal Service

Theoretical value: 1.00115965246
Experimental value: 1.00115965218

The difference between these two numbers is equivalent to the width of a hair in the distance between London and New York. In the entire history of physics no other prediction has been verified with this accuracy.

Richard Feynman

Thrush. Courtesy of An Post, Irish Post Office

When Feynman was a child in Far Rockaway, his father told him:

See that bird? It's a brown-throated thrush, but in Germany it's called a halsenflugel, in Portugese, it's a Untalaberdo, in Italian, it's a Chutapehda, in Chinese, it's a Chung-wa-tha and in Japanese, it's a Patathedahecha. Even if you know all those names for it, you still know nothing about the bird — you only know something about people; what they call that bird. Now that thrush sings, and teaches its young to fly, and flies so many miles away during the summer across the country, and nobody knows how it finds its way.

These words stayed with him always. He knew what it really meant to understand something and was determined to find things out for himself. Having made himself a little laboratory, he took apart all kinds of gadgets to learn how they worked and how to put them together again. He tinkered with electric motors, fixed radios and built an amplifier for a photocell; soon he acquired a reputation as the kid who could fix anything.

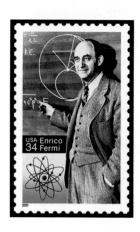

Enrico Fermi. Courtesy of the United States Postal Service

Young Feynman's interests were not confined to mechanical things. One time, he decided to study 'the mind of an ant'. He waited patiently until an ant walked onto a piece of paper, and then transported it to sugar some distance away. Would the ant find its way back to where it came from? Would it tell the other ants about the sugar? Would the ant lose its sense of direction if he rotated the paper before he put it down?

When he graduated from MIT, he decided to go to Princeton to find out about 'the rest of the world'. Princeton was quite different from other American universities. It was more like Oxford or Cambridge with 'a certain aspect of elegance'. On his first day at a tea party hosted by the dean's wife he heard her ask: '*Would you like cream or lemon in your tea Mr Feynman?*' Feynman, who had never before been at a tea party, was a bit flustered: '*I would like both please*', he heard himself say. '*Heh-heh-heh, surely you are joking Mr. Feynman*', she replied. Her reply became the title of a book of stories about his life.[7]

One day at Princeton, Bob Wilson came into Feynman's office and told him they had been funded to carry out a project involving the separation of different isotopes of uranium. Somewhat reluctantly, Feynman took on the job.

They were to go to Los Alamos in New Mexico and give theoretical back-up to the teams designing equipment for *The Manhatten Project*, a highly secret mission to build an atomic bomb. The project employed the best scientists that the United States could recruit.

When Feynman ultimately went to Los Alamos in April 1943, he had the opportunity to meet people such as Enrico Fermi (1901–1954), who had built the first nuclear reactor at the University of Chicago in 1942, Hans Bethe (1906–2005), who was then head of the theoretical division at Los Alamos, and the Hungarian-born mathematician, John von Neumann

John von Neumann. Courtesy of the United States Postal Service

(1903–1957). These men were 'household names' in science and mathematics.

The great man himself, Niels Bohr, arrived with his son Aage. They had travelled secretly from Denmark via Sweden and England. To preserve his secrecy, Bohr was given the pseudonym Nickolas Baker and became known as 'Uncle Nick'.

One morning Feynman received a phone call to say that Bohr wanted to see him. Feynman didn't think Bohr even knew he

---

[7] Richard P. Feynman. *Surely You're Joking Mr Feynman*. Norton & Company, New York. 1985.

existed, so why would he want to see a young student like himself? When he went to Bohr's office he found that Bohr wanted to discuss a new idea. They did just that for about two hours, until finally Bohr said: '*Well, I guess we can call in the big shots now*'.

Later, Aage told Feynman that his father had asked him to call that little fellow who had been sitting in the back row, saying: '*He is the only fellow who is not afraid of me, and will say if I have got a crazy idea*'. It was the same criterion that he had used to select young fresh minds for his Institute in Copenhagen!

All documents at Los Alamos were kept under lock and key. Having tinkered with locks as a boy, Feynman knew something about them and soon discovered that he could open the safe in his own office without knowing the combination. Nobody paid much attention when he tried to point out that the locks were unsafe; eventually he decided to draw attention to this lapse in security by opening a safe which contained classified material on the production of plutonium and purification procedures. Then he wrote a note saying: '*I borrowed document no. LA4312 — Feynman the safecracker*', and put it on top of the papers.

The escapade seemed to cause less alarm than might have been expected. Someone would say: 'Hey Feynman, my secretary is out and I have forgotten the combination, would you ever open the safe for me?' Other than that, the safes remained as they were. Once, on a visit to Oak Ridge, he showed a colonel that his safe could be opened. The next day a note was circulated to everyone in the plant:

If during his last visit here Mr Feynman was at any time in your office, near your office, or walking through your office, please change the combination of your safe.

Challenger disaster. Courtesy of NASA

On 28 January 1983 the space shuttle *Challenger* exploded, 73 seconds after its launch at Cape Canaveral in Florida and, tragically, all seven astronauts lost their lives. A few days afterwards Feynman received a phone call from William

Graham, the head of NASA, who introduced himself as a former student and asked Feynman to serve on an investigatory commission, to try to find the cause of the disaster. After some thought, Feynman agreed to serve on this commission even though by that time he was seriously ill with cancer.

In between the sittings of the commission, Feynman set out on his own to talk to the people from the ground up; technicians, mechanics and engineers. He did what he had done as a boy in Far Rockaway, find out for himself how things were put together and how they worked. He travelled alone to interview engineers at the Kennedy Space centre in Florida, the Marshall Space Flight Centre in Alabama and Johnson Space Centre in Houston. In between he had to make regular hospital visits to Washington, but was determined to complete the task which he had undertaken.

It soon became clear that the explosion could be traced back to the failure of the rubber in the '*O-rings*' which sealed the joints between the rocket and the fuel tanks. The temperature in Florida on the day of the launch was unusually low which critically affected the flexibility of the rubber and Feynman felt that the concerns of engineers regarding the launch on that day had not been given sufficient consideration.

In order to keep the public informed, the press were invited to some of the meetings of the commission. At one such meeting Feynman produced a sample of the rubber from the O-ring and dipped it into a glass of iced water. Using a small clamp he demonstrated that, for a few seconds at least, the material has no resilience when it is at a temperature of 32°F. He concluded: '*This, I believe, has some significance for our problem*'.

Feynman insisted that the final report of the commission contain an appendix which gave a number of his 'personal observations'. It finished with the observation: '*For a successful technology, reality must take precedence over public relations, for Nature cannot be fooled.*'

Feynman viewed the world with a unique insight and was driven by a determination to discover the essence of things, how they work, how they are put together: the mind of an ant, the mechanism of a lock, the design of a spacecraft.

Chapter 15

Quantum Reality — The World of the Absurd

Immaterialism

If a tree falls in a forest on an uninhabited island, is there a sound? This kind of question falls into the domain of *Immaterialism*, the philosophy of George Berkeley (1685–1753), which states that material things do not have an independent existence but exist only as mental perceptions or ideas; we live in a virtual world which only becomes real when we interact with it.

Thus, for example, the argument that the sound does not exist if it is not perceived might go as follows: 'Sound is a vibration, transmitted to our senses through the mechanism of the ear, and recognized as sound only at our nerve centres. The falling of the tree, or any other disturbance, will produce vibration of the air but if there are no ears to hear, there will be no sound.'

Berkeley was accomplished in many spheres. He graduated in mathematics and philosophy at Trinity College Dublin and then became a tutor and lecturer in Greek. When later he was made a bishop of the

George Berkeley. Courtesy of An Post, Irish Post Office

213

Anglican Church, he spent some time, also at Trinity, lecturing Divinity and Hebrew. The university of California and the town around it are named in his honour.

Reality and Perception

Berkeley argued that one cannot talk about an object *being*, only about it *being perceived*. The philosophy extended beyond things unheard or unseen to all physical phenomena. Attributes such as size, shape and colour are not absolute but depend on their perception. The whole universe depends for its existence on being perceived.

In his book *A Treatise Concerning the Principles of Human Knowledge*, published in 1710, Berkeley states:

> *It is agreed on all hands, that the qualities or modes of things do never really exist each of them apart by itself, and separated from all others, but are mixed, as it were, and blended together, several in the same object.*

With hindsight, it is amazing how closely this idea resembles the quantum mechanical concept of superposition of eigenstates.

Immaterialism was too radical for most of Berkeley's contemporaries. His ideas even attracted attention from the literary world; Samuel Johnson (1709–1784) is said to have commented '*I refute it thus*' and illustrated his response by kicking a rock with his shoe. How this act constituted a refutation is not clear. Subsequently, Johnson changed his mind and was 'converted' to immaterialism; he and Berkeley became good friends. More than a century later, James Joyce echoed the same sentiments in his epic novel *Ulysses* through *Stephen Dedalus* who is walking on the beach: '*Stephen closed his eyes to hear his boots crush crackling wrack and shells ... the ineluctable modality of the audible.*'

The Einstein–Podolsky–Rosen (EPR) Paradox

By the early 1930s, quantum mechanics was well established as the best, and seemingly the only, theory to describe phenomena on the atomic scale. There was no denying that it could successfully interpret all the experimental results which were then available and it had made many verifiable predictions.

Although he had played a central part in the original development of quantum theory, Einstein became more and more disenchanted as implications emerged which, to him, were unacceptable. He was convinced that the result of measuring a physical quantity is, in principle, absolutely predictable and does not depend on chance.

> *Quantum mechanics is certainly imposing. But an inner voice tells me that it is not yet the real thing. The theory says a lot, but does not really bring us any closer to the secret of the 'old one'. I, at any rate, am convinced that he is not playing at dice.*[1]

Einstein believed in the *reality* of the physical world, that it exists independently of whether or not it is observed: '*I like to think the moon is there even if I am not looking at it*'.

In 1935, Einstein and his research students, Boris Podolsky (1896–1966) and Nathan Rosen (1909–1995), at the Institute for Advanced Study in Princeton, New Jersey, published a paper entitled *Can Quantum-Mechanical description of Physical Reality be considered complete?* In this paper, the authors make a powerful attack on quantum theory, specifically on the assertion that physical quantities have no 'intrinsic' reality, but only exist when they are observed.

They propose what they feel to be a reasonable criterion of physical reality as follows:

> *If without in any way disturbing the system, we can predict with certainty (i.e. with probability equal to unity) the value of a physical quantity, then there exists an element of physical reality corresponding to that physical quantity.*

The basic tenet of quantum mechanics is that one cannot predict with certainty the value of any unmeasured physical observable. Neither the quantum mechanical description of the photon, nor of

---

[1] Letter from Einstein to Born. 4 December 1926.

the neutral kaon, fits that 'reasonable' criterion of reality. Whatever about the kaon, which had not been discovered in 1935, light is without doubt one of the central physical entities in the universe. If light is not real, we have a major problem!

Einstein and his co-authors consider a physical system, composed of two interacting parts, described by a common quantum mechanical wave function. According to the rules of quantum mechanics, if the two parts separate, any measurement of one part will change the wave function for the whole system. This will have an instantaneous effect on the other part, regardless of how far apart they are when that measurement is made.

Violating the Uncertainty Principle

The '*EPR*' paper makes a very general argument and then applies it to a specific thought experiment involving two particles, A and B, which interact and then separate, flying away from one another with equal and opposite momenta. The diagram below depicts the state of affairs some time later, when the particles have become separated.

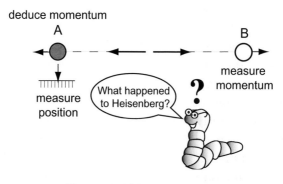

Einstein–Podolsky–Rosen paradox

By *measuring* the momentum of particle B, we can *deduce* the momentum of particle A without disturbing it in any way. If we now measure the position of A, we will have *exact* knowledge of both its momentum and position simultaneously, which violates Heisenberg's uncertainty principle, the foundation of quantum theory.

The only solution seems to be that we should go back to the classical idea that particle A must have had a definite position and a definite momentum all along.

The EPR paper concludes with the statement:

> *While we have shown that the wave function does not provide a complete description of the physical reality, we left open the question of whether or not such a description exists. We believe, however, that such a [complete] theory is possible.*

Another Thought Experiment

It is just not practicable to measure momentum and position with sufficient accuracy to test the uncertainty principle.

We can however imagine another experiment where the measurements, by their very nature, are easier to make. We will consider two electrons, bound together

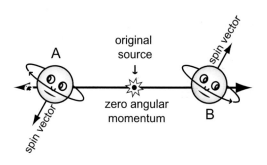

A thought experiment with spinning electrons

in a state which has a total angular momentum of zero.

If the electrons fly apart, the total wave function of the separated particles remains the same as when they were together. This means they have equal and opposite spins no matter how far apart they are.

In a quantum mechanical system, nothing exists until it is measured. Before the measurement, each electron is in a 50–50 mixture of *spin up* and *spin down* states. If we now measure the spin orientation of one electron and find it is in a spin up state, the bizarre prediction is that the other electron, no matter how far away it is, will instantly jump into a spin down state (and conversely). By making a measurement on the 'B electron', we instantly change the reality also of the 'A electron'. From the point of view of quantum mechanics the electrons are correlated, or *entangled*, in a bond which is independent of space and time.

By virtue of their spin, the electrons behave like tiny magnets and tend to align themselves in a magnetic field. It is then possible to separate them according to the orientation of their spins by sending them through an inhomogeneous magnetic field.[2]

'Spooky Action at a Distance'?

Einstein believed in 'local realism', which means that objects cannot be instantly affected by distant events and that their properties exist independently of any measurement. Physics should represent a reality in time and space free from '*spooky actions at a distance*' ('*Spukhafte Fernwirkungen*'). In the thought experiment, the spin property of the electron at A must have existed before any measurement was made. Einstein did not conclude that quantum mechanics was inherently wrong, merely that it was incomplete. Neither did he advocate a return to the comfort zone of classical physics where, whether we perceived it or not, the spins were there all the time.

Entangled Photons

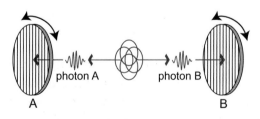

Entangled photons

Another step from thought towards practicality involves entangled photons whose polarization is the defining property, in place of the spin of electrons.

---

[2] Otto Stern (1888–1969) and Walther Gerlach (1889–1979) directly observed such separation in 1922.

Polarization lends itself to experiments which give YES/NO answers, equivalent to the UP/DOWN answers in the case of spin.

In the diagram above, A and B represent correlated photons which set off in opposite directions from a source situated between two polarizers. According to quantum mechanics, the polarization of the photons is indeterminate while they are in flight; whether either of the photons will pass or fail to get through its polarizer is decided by a symbolic throw of the dice. Even when the correlated photons are far apart, they must behave according to the same throw of the same dice.

Suppose that the two photons are polarized along the same axis and that the polarizers are aligned with their optical axes in the vertical direction. If the 'B photon' arrives first and is transmitted through the B polarizer, it will make a quantum jump into an *eigenstate* of vertical polarization. As a result of this quantum jump by B, its *entangled twin* A will pass through the A polarizer at the other end. Reality has changed for A, as if it had received a telepathic message from B: '*Have arrived and passed at 0^0. Be informed you are vertically polarized and will pass at 0^0*'. The same sort of thing happens if the situation is reversed and the A photon arrives first.

Hidden Variables

Unpredictable quantum jumps, telepathic messages and spooky actions at a distance make good arguments against the credibility of quantum theory.

The most obvious solution is that *hidden variables* govern quantum processes and they are the means of making quantum theory complete. These variables would follow the rules of classical determinism but are unknown, at least for now.

The idea of hidden variables is attractive because they are commonplace in the classical

Lottery draw

world. Many events in the household world seem to occur apparently at random and look like the quantum jumps of the atomic world. An example of such a 'pseudo quantum jump' is an event like winning a lottery. One moment there are a million people, each holding a ticket worth one dollar; at the next instant, they all hold a worthless ticket except one person, who has made a quantum jump and become a millionaire. This looks like a chance event that nobody could have predicted; any sort of predetermined lottery would constitute a major fraud.

The Hidden Variables in the Lottery

Theoretically, a lottery draw can be considered predictable, without being fraudulent, because no one had complete information about the hidden variables, or what Richard Feynman called 'hidden gears and wheels', behind the quantum jump. In the lottery, the hidden variables are coloured balls with different numbers, which are blown around in a transparent cage. One by one, they drop into a tube to form the winning lottery number.

In theory, if we knew the position and momentum of every single molecule in the mechanism of the system, classical mechanics would allow us to predict the winning number. Classical mechanics is deterministic and nothing occurs by chance, but in such a complicated situation, what we can do in principle cannot be done in practice. Events with so many variables are beyond the capabilities of computation.

Hidden Variables in the Computer

There are hidden variables associated with so-called random number generators. Computer methods of producing random numbers are based on mathematical algorithms of greater or lesser complexity, but every algorithm must be given a 'seed number' to start the process of computation. A particular algorithm will generate exactly the same sequence of 'random numbers' every time the same seed number is used. This means that these numbers are not really random. The seed number is the hidden variable, the key to that particular code.

Nothing is truly random in the household world, not even the best made dice, not to mention the loaded ones! There are many other ways of getting *pseudo random* distributions, such as shuffling cards and spinning roulette wheels.

In the household world there are always hidden variables.

The Case for Hidden Variables in Quantum Theory

It is not surprising that the dismissal of hidden variables by Bohr, and other supporters of the Copenhagen interpretation, was questioned by Einstein, particularly since the alternatives were so bizarre. Perhaps, deep inside the electron, proton and even the photon, there are hidden gears and wheels which, if we are clever enough, we may discover some day.

Accordingly, each photon would have a 'list of instructions' as to what action to take in any given situation. It would be immaterial whether the origins of the instructions, or the reasons for them, were known or understood. The properties of the photon would be independent of any observer and its behaviour entirely predictable. Correlated photons would carry identical 'instruction manuals' telling them, for example, whether or not to pass through a polarizer inclined at any particular angle. There would be no need for telepathic communication.

The Case for a Quantum Theory Without Hidden Variables

Quantum theory had given an explanation for most atomic and subatomic phenomena and had made many predictions, which were subsequently verified in minute detail. The evidence that the theory is built on solid and correct foundations was overwhelming. These foundations were largely conceived at Bohr's school in Copenhagen and, according to that school, specifically exclude hidden 'gears and wheels'.

Further apparent confirmation came in 1932, when the Hungarian-born mathematician, János von Neumann (1903–1957), published a book called *The Mathematical Foundations of Quantum Mechanics*,[3] in which he put quantum theory on a firm mathematical basis. The book contained a proof that hidden variables are incompatible with quantum mechanics. This strengthened the Copenhagen argument and stifled the search for an 'ordinary' reality beneath quantum theory.

Von Neumann's Blind Spot

Von Neumann was acknowledged as one of the great mathematicians of the day and most physicists and mathematicians accepted the hidden variables proof without question. One exception was Grete Hermann (1901–1984) a German mathematician who, three years after von Neumann's book was published, made the assertion that the proof was invalid. Her claim was largely ignored.

David Bohm (1917–1992) also believed von Neumann's argument was wrong but was unable to find the flaw. In 1952, he developed what is arguably the best known hidden variable theory; however, it includes action at a distance, the very thing Einstein was anxious to avoid.

When All Else Fails Can We Fall Back on Hidden Variables?

The question of hidden variables remained unresolved. Are hidden variables something we can hold in reserve, as some sort of insurance against possible failures of quantum theory, without saying what the variables are or

---

[3] János von Neumann. *The Mathematical Foundations of Quantum Mechanics*. Princeton University Press. Princeton NJ. 1996.

how to find them? Leaving aside the probabilistic interpretations of quantum mechanics, is it possible to devise a hidden variables theory to fit any given set of experimental results?

Enter John Bell

John Stewart Bell (1928–1990), was employed in 1944 as a technician in a physics laboratory at Queen's University Belfast, Northern Ireland. His talent and interest in physics immediately became obvious to the lecturing staff, who invited him to attend first-year lectures while still working as a technician. With savings

John Stewart Bell. Courtesy of Queen's University, Belfast

from his salary he was able to enter the university as a student in 1945, the first step in an illustrious career.

In 1960, Bell obtained a position at the accelerator laboratory at CERN, where he worked on theoretical particle physics and also on accelerator design. He carried out his 'day job' meticulously. His spare time was devoted to the development of a deeper understanding of the essence of quantum mechanics. He wanted to throw some light on the issues debated by Bohr and Einstein and, in particular, to find an answer to the question of hidden variables. Ideally, philosophical issues would then be converted into scientific questions which could be answered by experiment.

In 1964, Bell published a paper[4] on the EPR paradox in *Physics*, a rather obscure and short-lived journal. The paper contained a theorem, which has been described as one of the most influential theorems in quantum mechanics. The importance of Bell's theorem was not recognized at the time, and it was more or less ignored for about five years. Then came the realization that it provided a mathematical solution to what had been treated as a philosophical question.

---

[4] J.S. Bell. *On the Einstein–Podolsky–Rosen Paradox*. Physics 1, 195–200 (1964).

Bertlmann's Socks and the Nature of Reality

Shortly before his premature death in 1990, Bell gave a lecture at University College Dublin[5] in which he described his theorem in simple terms, accessible to a non-specialist audience. Having explained the conundrum posed by the EPR paradox, he recalled how at CERN he had worked with Reinhold A. Bertlmann from the University of Vienna, who had a somewhat unusual habit. As a token of dissent from irrational conventions Bertlmann always wore non-matching socks of different colour.

Which colour he will wear is quite unpredictable but when he comes round the corner and you see that one of his socks is pink you can be sure that the other sock is of some colour other than pink.

Bertlmann's odd sock

We might add other bits of information which instantly become available; for example, there is a high probability that there is an odd pink sock left in a drawer in Bertlmann's apartment. This does not mean that our observation has made a dramatic change to reality around the corner or at some distant apartment. The hidden sock was there all the time.

To quote Bell:

> *The philosopher in the street, who has not suffered a course in quantum mechanics, remains singularly unimpressed by Einstein–Podolsky–Rosen correlations. He can point to many examples of similar correlations in everyday life ... there is no accounting for tastes, but apart from that there is no mystery here. And is not the EPR business just the same?*

Why can an equally trivial solution not apply to the EPR paradox?

---

[5] Some quotations from this talk are given from memory and are not verbatim.

Genetic Predisposition

Bell illustrated his talk with a story involving identical twins who have unusual 'hidden variables' in their genes. In this piece of fiction, the twins are tested for genetic predisposition to certain foods, specifically fatal allergies to apples, bananas and cabbage. (In his talk Bell used hamburger, pizza and spaghetti, but we have changed the foods for purposes of notation.)

His reason for using twins in the tests was that, should one twin succumb to a particular food, the surviving twin can still be tested for one of the other allergies. Each twin is tested for just two of the three foods.

The results are then compiled using the notation:

$$A^+ \rightarrow \text{lives on apple} \qquad A^- \rightarrow \text{dies on apple}$$
$$B^+ \rightarrow \text{lives on banana} \qquad B^- \rightarrow \text{dies on banana}$$
$$C^+ \rightarrow \text{lives on cabbage} \qquad C^- \rightarrow \text{dies on cabbage}$$

Bell then made the following statement which sounds simple:

The number of twins who live on apple but die on banana plus the number who live on banana but die on cabbage is greater or equal to the number who live on apple but die on cabbage.

In case the statement was not immediately obvious, which was certainly true for the authors and for the majority of the audience, Bell then provided the following simple proof.

We can define three groups:

$N(A^+B^-)$ the number of twins who live on apple but die on banana,

$N(A^+C^-)$ the number of twins who live on apple but die on cabbage and

$N(B^+C^-)$ the number of twins who live on banana but die on cabbage;

These three groups can be subdivided according to their reaction to the remaining food.

Some of the A^+C^- group also die on banana; the remainder live on banana i.e.

$$N(A^+ C^-) = N(A^+ B^- C^-) + N(A^+ B^+ C^-).$$

Similarly, the A^+B^- group can be divided according to their reaction to cabbage i.e.

$$N(A^+ B^-) = N(A^+ B^- C^+) + N(A^+ B^- C^-)$$

which means that $N(A^+ B^-) \geq N(A^+ B^- C^-)$.

And finally the B^+C^- group can be divided according to their reaction to apple i.e.

$$N(B^+ C^-) = N(A^+ B^+ C^-) + N(A^- B^+ C^-)$$

which means that $N(B^+ C^-) \geq N(A^+ B^+ C^-)$.

If we now add $N(A^+ B^-)$ and $N(B^+ C^-)$:

$$N(A^+ B^-) + N(B^+ C^-) \geq N(A^+ B^- C^-) + N(A^+ B^+ C^-) = N(A^+ C^-)$$

$$\boxed{N(A^+ B^-) + N(B^+ C^-) \geq N(A^+ C^-) \text{ Bell's inequality.}}$$

This is a simplified statement[6] of Bell's inequality, an algebraic relation involving elements which have an independent value, i.e. satisfy Einstein's criterion of reality and are immune from spooky actions at a distance (are *realistic* and *local* in scientific terms).

In the above story, Bell's inequality will be satisfied if the allergies are coded in the genes of the twins. We do not have to know the code or where to find the key; if there are such *hidden variables*, they constitute fixed properties and the results of any tests will be consistent with the inequality. Bell's inequality applies to any test which requires a YES or NO answer such as whether a twin survives on a certain food, or a photon passes through a polarizer.

The extraordinary significance of Bell's formula was succinctly described in a remark made in 1982 by Richard Feynman: '*It seems*

---

[6] Bell himself refers to this as the *Wigner–d'Espagnat relation,* which serves as an introduction to a completely general statement which bears Bell's name.

almost ridiculous that you can squeeze it (the difficulty of quantum mechanics) *into a numerical question that one thing is bigger than another'*.

A Simple Thought Experiment

$$A^+B^- + B^+C^- \geqslant A^+C^-$$

Result cards

Let us try to devise a set of hidden variables in the form of 'instructions', that do not satisfy the inequality.

We will simulate a real life situation by recruiting volunteers who will each be given three tasks (A, B and C) which they will either pass or fail.

As we start to prepare 'result cards', we realize that we can take a shortcut by preparing only cards which already contain the instruction 'PASS A, FAIL C', corresponding to the right-hand side of the inequality. Much to the relief of the volunteers, they need only attempt task B.

Every card in the box on the right also qualifies for the box on the left

We take two boxes and label the right-hand box 'A$^+$ C$^-$' and the left-hand box 'A$^+$ B$^-$ + B$^+$ C$^-$'. The volunteers go off to attempt task B and, as they return, we can start to fill in the cards. We will disprove the inequality if we end up with more cards in the right-hand box than in the left-hand box.

The rationale behind the inequality becomes apparent immediately. If the first volunteer passes task B, we put a √ for PASS opposite B. This creates the instruction $A^+ B^+ C^-$, which means the card can be put into either box.

Similarly, if another volunteer fails the task and we put an X for failure opposite B, this creates the instruction $A^+ B^- C^-$. This card also can be put into either box.

We can make out a duplicate card each time and put one card into each box, but the fact remains:

Every card which goes into the box on the right will also qualify for the box on the left.

Any additional combinations of instructions will either go in the left-hand box or not enter the equation. We must apologize to the volunteers and send them home. We can throw away the rest of the cards.

Bell was quite right when he said his statement is obvious. Every twin, or for that matter every individual, who will live on apple but die on cabbage will either survive or succumb to a meal of bananas. He or she, therefore, also qualifies for the left-hand side of the equation. In addition there will be some who live on cabbage and some who die on apple who would go only on the left-hand side, which changes the equation into an inequality.

Bell's inequality applies to any property or value which is *real*, i.e. exists independently of the observer. To paraphrase Einstein's definition of a property that is real: '*If we can predict with certainty the value without disturbing it in any way*'. The property does not have to be hidden; the only reason we refer to 'hidden variables' is that we don't know what they are.

The twins in Bell's example, despite their peculiarities, are classical household entities obeying realistic laws and immune from actions at a distance. If they have allergies, the results of tests will conform with Bell's inequality. It remains to be seen if this is the case for photons.

From Twins to Photons

The analogy can be translated from twins to photons and from allergies to the orientations of polarizers. We can then test by experiment if the results are consistent with properties which have a physical reality.

In the case of the 'allergic' twins there are three instructions: how to react to apple, banana and cabbage. In the case of photons, we might start with a simple list of whether to pass or fail for three settings of the optical axis, say 0°, 30° and 60°. We do not have to understand why we are giving these instructions.

Using the same notation as previously, Bell's inequality reads:

$$N(\text{Pass at } 0, \text{fail at } 30) + N(\text{Pass at } 30, \text{fail at } 60)$$
$$\geq N(\text{Pass at } 0, \text{fail at } 60).$$

As we have seen, quantum mechanics predicts that the probability that photons will pass through two consecutive polarizers with axes inclined at an angle θ to one another is $\cos^2\theta$; the corresponding probability for failure is $(1 - \cos^2\theta)$. This result, Malus' law, has been well established experimentally.

Using Malus' law, we can calculate the numbers. We find that:

$N(\text{Pass at } 0, \text{fail at } 30) = \cos^2 30 = 0.25$
$N(\text{Pass at } 30, \text{fail at } 60) = \cos^2 30 = 0.25$, so that
$N(\text{Pass at } 0, \text{fail at } 30) + N(\text{Pass at } 30, \text{fail at } 60) = 0.5$.

However, $N(\text{Pass at } 0, \text{fail at } 60) = \cos^2 60 = 0.75$ (which is larger than 0.5!).

In terms of the more familiar food allergies, we have shown that:

According to quantum mechanics:
$$N(A+B-) + N(B+C-) < N(A+C-);$$

whereas, according to Bell's inequality,
$$N(A+B-) + N(B+C-) \geq N(A+C-).$$

This means that quantum mechanics is inconsistent with Bell's inequality and therefore with theories based on locality and realism. The question 'Which theory correctly describes the laws of nature?' can be answered only by experiment.

From Thought to Action, Experimental Tests of Bell's Theorem

An experimental test of Bell's theorem requires large numbers of correlated particles. Photons are easier to produce in the laboratory than electrons so the experiment with polarized photons offers the best chance of success.

An Idealized Experiment Using Entangled Photons

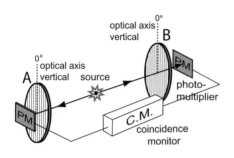

Idealized experiment with entangled photons

We will consider a real, but idealized experiment to test Bell's inequality. Correlated photons are emitted from a source between two polarizers, as shown in the diagram. Photomutiplier detectors behind the polarizers record the arrival of photons. If either detector is activated, it sends an electrical pulse to a *coincidence monitor* which registers a count every time pulses from A and B arrive simultaneously.[7]

Data is classified under two headings. A coincidence MATCH is recorded when the entangled photons both pass through their respective polarizers. A MISMATCH is recorded when one photon is transmitted while its partner at the other end is absorbed.

In this idealized set-up, the polarizers, photomultipliers and counters are 100% efficient; there are no stray photons, there is no background noise and the photomultipliers are activated by a single photon.

Parallel Settings

First we confirm that the separated photons have remained entangled, by setting the optical axes of both polarizers in the same direction (say,

---

[7] We will assume that allowance is made for any minute time lag due to difference in distances from the source S to the counters at B and A.

vertical). Entangled photons should act as one, and always either pass together or fail together[8] when the settings of the polarizers are parallel (whether they are set to 0°, 30° or any other angle).

Two Simultaneous Measurements

By making measurements on a photon and its entangled twin, we are effectively making two measurements on the same photon, knowing for sure that the first measurement does not disturb the other. Suppose we rotate the axis of the A polarizer through 30°, while keeping the axis of the B polarizer at 0°. This will give us information on whether the photon passes or fails at the two settings of 0° and 30°. We can repeat this process as many times as we like, using other photons and other pairs of angles.

We can now start to gather statistics. For example, out of a million photons which pass at 0°, how many pass or fail at 30°, 60° or any other angle, for that matter.

In principle, it is relatively straightforward to devise a method of testing Bell's inequality using photons. However, there are considerable technical difficulties associated with the actual experiment. The requirements are numerous; a plentiful supply of correlated photons, efficient photon detectors, efficient polarization, efficient counting techniques and efficient coincidence monitors — a formidable 'wish list'.

In practice nothing is ideal, the detectors may not respond or may be triggered by stray light. Many apparent mismatches will be caused by either of these effects.

An Actual Experiment Using Entangled Photons

In 1974, Alain Aspect, a graduate student at the *Institut d'Optique Théorique et Appliquée* at the University of Paris, began to plan a

---

[8] Strictly speaking, since the polarization of entangled photons is mutually perpendicular, each always does the exact opposite of the other. To avoid unnecessary confusion we will assume that this is accounted for 'behind the scenes' by a 90° adjustment to the angular calibration of one polarizer.

Source of entangled photons 1981.
Courtesy of Alain Aspect, Institut
d'Optique, Palaiseau, France

comprehensive series of experiments to test Bell's theorem. He went to see Bell to talk about his plans, but Bell was not very optimistic. The first question Bell asked was: '*Do you have a permanent position?*' On hearing that this was not the case, he described Aspect as a very courageous graduate student.

The efficiency of photomultipliers as they are used in such experiments is so low that the only correlation directly measurable is MATCH, where both photomultipliers record a hit. When only one photomultiplier records a hit, there is ambiguity as to whether the other photon was blocked by the polarizer or simply missed by the detector, so the measurements are incomplete.

In 1982, Aspect and his collaborators published the results of an experiment[9] in which it was possible to measure (rather than infer) coincidence rates for *all* possible combinations. The conventional polarizers were replaced by '*polarizing cubes*' which transmit both parallel and perpendicular polarizations. Photons polarized parallel to

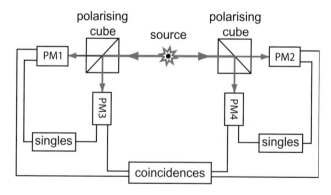

Coincidence counting in the Aspect experiment.

[9] A. Aspect, J. Dalibard and G. Roger. *Experimental Test of Bell's Inequalities Using Time-Varying Analyzers.* Physical Review Letters 49, 1804–1807 (1982).

the optical axis pass straight through the cubes, and enter either PM1 or PM2. Photons polarized perpendicular to the optical axis are reflected at the junction of the prisms and enter either PM3 or PM4.

The illustration above shows the electrical circuit used to count coincidences and single triggers of the photomultipliers at each end of the apparatus.

The left- and right-hand systems were mounted on a rotatable mechanism and measurements were made over the entire 360° range of relative orientations of the polarizers.

The Result of the Experiment

The experimental results are summarized in the graph which shows the coincidence rate as a function of the angle θ between the optical axes of the two polarizers. The experimental points are plotted on the background of the $\cos^2\theta$ curve representing the quantum mechanical pre-diction. The agreement is

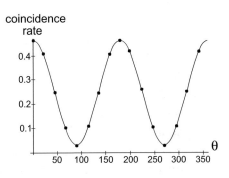

Quantum mechanics is right!

uncanny; the experiment agrees with quantum mechanics and violates Bell's theorem.

Closing all Possible Loopholes

A result which contradicts Bell's inequality cannot be accepted lightly and a number of 'loopholes', which might invalidate the experimental result, were suggested. One was that, due to the inefficiency of the instruments, the measurements were made on a biased sample of photons. No particular reason was given for such a bias. Bell's own comment was:

> *It is hard for me to believe that quantum mechanics works so nicely for inefficient practical set-ups and is yet going to fail badly when sufficient refinements are made.*

However, Bell thought that it would be good to test another, equally unlikely, suggestion. This was that, since the settings of the instruments were made in advance, the polarizers could reach some kind of mutual rapport by exchanging information 'at leisure' before the experiment. There would then be no need for signals at a speed greater than the speed of light. The interaction would then have existed, not between the photons, but between the polarizers.

Leaving no stone unturned, and to close every possible loophole, Aspect and his team then completed another experiment, in which the settings of the polarizers were changed while the photons were in flight. This was accomplished by electronic switching, which rapidly re-directed the light from one polarizer setting to another.

How Much Time Do We Have?

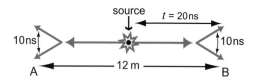

Settings switched while the photons are in flight

In this context, where the time intervals are extremely short, it is more appropriate to express time in *nanoseconds* (1 ns = one billionth of a second). The separation between the analysers in Aspect's experiment was 12 m, which means the photons took 20 ns to travel from the source to the detector. The 'optical switching' was done in about 10 ns, just about fast enough to change the settings while the photons were in transit.

The results were just the same as in the first experiment. Introducing the switching procedure made no difference. The results were as predicted by quantum mechanics in violation of Bell's inequality.

The Innsbruck Experiment

In 2008, Gregor Weihs and his team at the university of Innsbruck in Austria, published the results of an experiment in which they

University of Innsbruck

The Innsbruck experiment. Courtesy of Gregor Weihs, University of Innsbruck

used Aspect's method, with a number of improvements.[10] The entangled photons were sent to 'observation points' 400 m apart (on opposite sides of the science campus). This modification increased the 'time of flight' of photons travelling between the source and the detectors from 20 ns to 650 ns. The photons were guided by optical fibres and arrived at their respective destinations within 5 ns of each other.

[10] G. Weihs *et al.*, *Violation of Bell's inequality under strict Einstein locality conditions*, Physical Review Letters 81, 5039 (1998).

The measurements were made at what were essentially two independent 'mini-laboratories' occupied by observers 'Alice' and 'Bob'. Each 'laboratory' had its own random number generator, according to which the settings of the polarizers were changed many times while the photons were in flight.

To make sure there was no conceivable electronic connection between the measuring points, coincidences were not recorded by a common coincidence circuit, as in Aspect's experiment. Instead, each observer had his own atomic clock and registered the individual events independently. The numbers of events would be compared much later, after the measurement was finished.

Summary of the Facts

As each photon arrives at its destination, a polarization analyser gives it a test which the photon can PASS or FAIL. At the last moment (less than 100 ns from arrival), the photon is redirected towards a polarizer setting, selected randomly while both photons were in flight.

There was no communication between the mini-laboratories. In the event that a message was sent, even at the speed of light, it would take about 1300 ns to travel from one laboratory to the other.

When everything is finished, the observers get together and compare their records. They find that the photon twins always either pass together or fail together when the optical axes of the analysers are set in the same direction (MATCH).

When the optical axes are perpendicular, the observers find that results are opposite; if one photon fails, its twin at the other side of the campus passes and vice versa (MISMATCH).

When the optical axes point in different directions, the percentage of matches is proportional to $\cos^2\theta$ (where θ is the angle between the two axes) in agreement with quantum mechanics. The results do not satisfy Bell's inequality, which means that the outcome of the experiment cannot be explained by hidden variables.

This loophole, at least, seems to be tightly closed. The EPR paradox is not solved by hidden variables.

Epilogue

In Greek mythology, *Atlas* fought with the Titans in their war against the Olympians. When the Titans lost, Atlas was condemned to stand for ever at the edge of the Earth and to support the sky on his shoulders. The myth was not intended to be taken literally, and the philosophers could not claim to have the key to the 'mysteries of the heavens'.

Statue of Atlas. Collezione Farnese, Museo Archeologico Nazionale di Napoli

The fact remains that things fall unless they are supported. The sky has no visible means of support and yet it does not fall: the story of Atlas provided a symbolic explanation in terms which were in accordance with common experience.

Many centuries later, Isaac Newton presented a much better explanation. By combining his laws of motion and his theory of gravitation, he showed that the moon and other celestial bodies move in orbits which allow them to '*fall without getting nearer to one another*'. They execute a dynamic pattern which is stable and beautifully organized. Newton recognized immediately that he had found nature's way of dealing with the problem of the unsupported sky: '*All the mechanics of the Universe at once lay spread before me*'.

Newton's concepts of planetary motion and gravitational action at a distance were more difficult to visualize than the figure of Atlas carrying the celestial sphere. His ideas would have been unlikely to appeal to ancient philosophers, because they were contrary to common experience. Nature is not constrained by what we might consider to be reasonable. She does things in the best and most efficient manner, regardless of what we might think would be the best way.

When Albert Einstein developed the theory of relativity, he introduced a new concept of space and time. He wiped the slate clean of 'preconceived prejudices', bringing in new ideas which were counterintuitive and which many of his contemporaries considered to be absurd. Undeterred, Einstein stood by his theory because in it he saw a deeper and more logical view of the world. Just as Newton before him, he was convinced that his theory was right. There just could not be any other way. It had to be like that, otherwise '*God would have missed a wonderful opportunity*'.

As the quantum adventure progressed, the consequences moved further and further away from common experience. A universe controlled by the throw of a dice, reality which depends on being observed, 'spooky' actions at a distance; these were concepts too absurd, even for Einstein. Could it be that Einstein's slate was not sufficiently clean? As we push the limits of knowledge, are we seeing more examples that the world is different from what we expect it to be? If God does not throw dice, the pretence is perfect.

Milestones

Some Key Events in the Story

1900　October 19　Planck's 'inspired guess'. 'Comment' at a meeting of the German Physical Society. *On an improvement on Wien's radiation law,* Verhandlungen der Deutschen Physikalischen Gesellschaft 2, 202–204 (1900).

　　　　December 14　Planck's quantum hypothesis. Presented at a meeting of the German Physical Society. *On the theory of the energy distribution law of the normal spectrum,* Annalen der Physik 4, 553–563 (1901).

1905　March 17　Einstein's paper on the photoelectric effect. *On a heuristic viewpoint concerning the production and transformation of light,* Annalen der Physik 17, 132–148 (1905).

　　　　September 27　Einstein's relativistic equation $E = mc^2$. *Does the inertia of a body depend on its energy content?* Annalen der Physik 18, 639–641 (1905).

1916　March　Millikan's photoelectric experiment confirms value of Planck's constant. *Direct measurement of Planck's constant h,* Physical Review 7, 355–388 (1916).

1923   November   Compton Effect. *The spectrum of scattered X-rays*, Physical Review 22, 409–413 (1923).

1924   November 29   De Broglie submits doctoral thesis proposing wave properties for articles. *Recherches sur la théorie des quanta* (Researches on the quantum theory), Annales de Physique (Paris) 3, 22 (1925).

1925   July 29   Heisenberg's first paper (submitted by Born). *Quantum-theoretical reinterpretation of kinematic and mechanical relations*, Zeitschrift für Physik 33, 879–893 (1925).

       September 25   Born and Jordan. *On quantum mechanics*, Zeitschrift für Physik 34, 858–888 (1925).

       November 16   Born, Heisenberg, Jordan. *On quantum mechanics II*, Zeitschrift für Physik 35, 557–615 (1925).

       December   Dirac. *The fundamental equations of quantum mechanics*, Proceedings of the Royal Society of London A 109, 642–653 (1925). (Submitted nine days before the Born–Heisenberg–Jordan paper appeared in Zeitschrift für Physik).

1926   January   Schrödinger equation. *Quantization as an eigenvalue problem*, Annalen der Physik 79 (4), 361–376 (1926).

       May   Schrödinger shows wave mechanics and matrix mechanics are equivalent. *The relationship of the Heisenberg–Jordan quantum mechanics to mine*, Annalen der Physik 79 (4), 734–756 (1933).

       July   Schrödinger solution of hydrogen atom. *Quantization as an eigenvalue problem (third communication)*, Annalen der Physik 80 (13), 437–490 (1926).

1927		Davisson and Germer confirm wave nature of electron. *Diffraction of electrons by a crystal of nickel,* Physical Review 30 (6), 705–740 (1927).
1928		Dirac's equation gives negative energy solution. (At first Dirac 'does not dare' postulate a new particle). *The quantum theory of the electron,* Proceedings of the Royal Society of London A 117, 610–624 (1928).
1933	March 15	Anderson observes positive electron. *The positive electron,* Physical Review 43, 491–494 (1933).
1934		Yukawa proposes the existence of new particle at a meeting of the Physico-Mathematical Society in Osaka in 1934. His lecture was published in the following year: *On the interaction of elementary particles I,* Proceedings of the Physico-Mathematical Society Japan 17, 48–57 (1935).
1935	May	Einstein–Podolski–Rosen paradox. *Can quantum-mechanical description of physical reality be considered complete?* Physical Review 47, 777–780 (1935).
1942		Feynman's Ph.D. thesis: *The principle of least action in quantum mechanics,* Princeton University.
1949	September	Feynman: Quantum electrodynamics. *Space-time approach to quantum electrodynamics,* Physical Review 76, 769–789 (1949).
1955		Gell-Mann and Pais propose neutral kaon as a long and short-lived two-state quantum mechanical system. *Behavior of neutral particles under charge conjugation,* Physical Review 97, 1387–1389 (1955).

Pais and Piccioni suggest experiment to demonstrate eigenstates of neutral kaon. *How to verify experimentally a recent theoretical suggestion that the K^0 meson is a particle mixture,* Physical Review 100, 1487 (1955).

1960   Piccioni and his team carry out the experiment. *Regeneration and mass difference of neutral K-mesons,* Physical Review Letters 4, 418 (1960).

1964   Bell's Theorem. *On the Einstein–Podolsky–Rosen paradox,* Physics 1, 195–200 (1964).

1982   Aspect experiment. *Experimental test of Bell's inequalities using time-varying analyzers,* Physical Review Letters 49, 1804–1807 (1982).

1998   Gregor Weihs and his team observe entangled photons on either side of the campus at the University of Innsbruck. *Violation of Bell's inequality under strict Einstein locality conditions,* Physical Review Letters 81, 5039 (1998).

Index